Routledge Revivals

Technology Options for Electricity Generation

Environmental constraints and market uncertainties create new challenges for electricity generation. In this title, originally published in 1991, the authors present a simulation model with a capability for highly detailed activity to identify cost-minimising investment options under different assumptions about demand, costs, regulation, and other economic and environmental factors. Applying the model to two U.S. regions having sharply different electricity demand and supply characteristics, they identify the importance of advanced technologies and augmented electricity trade among regions. This title is ideal for students interested in environmental studies.

Technology Options for Electricity Generation

Economic and Environmental Factors

Edited by
Hadi Dowlatabadi and Michael A.
Toman

RFF PRESS
RESOURCES FOR THE FUTURE

First published in 1991
by Resources for the Future, Inc.

This edition first published in 2016 by Routledge
2 Park Square, Milton Park, Abingdon, Oxon, OX14 4RN
and by Routledge
711 Third Avenue, New York, NY 10017

Routledge is an imprint of the Taylor & Francis Group, an informa business

Publisher's Note
The publisher has gone to great lengths to ensure the quality of this reprint but points out that some imperfections in the original copies may be apparent.

Disclaimer
The publisher has made every effort to trace copyright holders and welcomes correspondence from those they have been unable to contact.

A Library of Congress record exists under LC control number: 91006324

ISBN 13: 978-1-138-95914-9 (hbk)
ISBN 13: 978-1-315-66078-3 (ebk)
ISBN 13: 978-1-138-95915-6 (pbk)

TECHNOLOGY OPTIONS FOR ELECTRICITY GENERATION

ECONOMIC AND ENVIRONMENTAL FACTORS

Hadi Dowlatabadi and Michael A. Toman

Published by Resources for the Future, Washington, DC

Distributed by The Johns Hopkins University Press, Baltimore, MD

Printed in the United States of America

Published by Resources for the Future
1616 P Street, N.W., Washington, D.C. 20036
Books from Resources for the Future are distributed worldwide by
The Johns Hopkins University Press

Library of Congress Cataloging-in-Publication Data

Dowlatabadi, Hadi.
   Technology options for electricity generation : economic and
environmental factors / Hadi Dowlatabadi and Michael A. Toman.
     p. cm.
   Includes bibliographical references.
   ISBN 0-915707-58-6 (alk. paper)
   1. Electric utilities—Costs. 2. Electric power-plants—
Environmental aspects. I. Toman, Michael A. II. Title.
HD9685.A2D65 1991
333.79'32—dc20
                       91-6324
                       CIP

This book was edited by Samuel Allen and designed by Debra Naylor. The cover was designed by Gehle Design Associates.

∞ The paper in this book meets the guidelines for permanence and durability of the Committee on Production Guidelines for Book Longevity of the Council on Library Resources.

RESOURCES FOR THE FUTURE (RFF) is an independent nonprofit organization that advances research and public education in the development, conservation, and use of natural resources and in the quality of the environment. Established in 1952 with the cooperation of the Ford Foundation, it is supported by an endowment and by grants from foundations, government agencies, corporations, and individuals. Grants are accepted on the condition that RFF is solely responsible for the conduct of its research and the dissemination of its work to the public. The organization does not perform proprietary research.

RFF research is primarily social scientific, especially economic. It is concerned with the relationship of people to the natural environmental resources of land, water, and air; with the products and services derived from these basic resources; and with the effects of production and consumption on environmental quality and on human health and well-being. Grouped into four units—the Energy and Natural Resources Division, the Quality of the Environment Division, the National Center for Food and Agricultural Policy, and the Center for Risk Management—staff members pursue a wide variety of interests, including forest economics, natural gas policy, multiple use of public lands, mineral economics, air and water pollution, energy and national security, hazardous wastes, the economics of outer space, climate resources, and quantitative risk assessment. Resident staff members conduct most of the organization's work; a few others carry out research elsewhere under grants from RFF.

Resources for the Future takes responsibility for the selection of subjects for study and for the appointment of fellows, as well as for their freedom of inquiry. The views of RFF staff members and the interpretation and conclusions of RFF publications should not be attributed to Resources for the Future, its directors, or its officers. As an organization, RFF does not take positions on laws, policies, or events, nor does it lobby.

CONTENTS

LIST OF FIGURES

LIST OF TABLES

vii

PREFACE

This book reports the results of a project begun in the fall of 1987 as part of a new RFF research program on issues in the electricity generation industry. We undertook this project holding two basic convictions. The first of these is that understanding current economic and policy issues in electricity requires more than a grasp of the industry's economic structure and regulation, as important as these elements are. It also requires an analysis of the interplay between economic forces and the industry's path of technological evolution. Just as the development of different technologies has been and will continue to be significantly influenced by economic forces (including regulation), the evolution of generation technology alters (for the good, we believe) the industry's abilities to respond to these forces. An assessment of problems facing the electricity industry and their solutions which ignores these technological dimensions may be seriously misleading.

Our second conviction is that addressing these technological issues requires a detailed analytical framework. A simple engineering assessment of the characteristics of different technologies is unlikely to be sufficient. Economic analyses based on simple constructs, like an aggregate cost function, also may be uninformative or misleading given the unique features of electricity supply, as discussed in the classic monograph by Turvey (1968). Many frameworks combining the essential technological and economic elements have been developed for studying both the planning problems faced by individual utilities and issues of broader concern, such as the impact of changing environmental regulations on the costs and emissions of the industry. A common weakness of existing methodologies, however, is a limit on the ability to address the sensitivity of conclusions to different economic and technological influences. Such a capability is vital given the manifold uncertainties we face by gauging future influences on electric supply.

The research reported in this book reflects these convictions. We have sought to strengthen existing methodology for analyzing technology choice in electricity generation with uncertain futures. While much remains to be done in developing the framework, we believe the work reported here is a successful beginning.

In applying the model, we also have sought to address a fundamental question involving the interaction of economics and technology: What economic cost-minimizing investment decisions can be made under different assumptions about demand, costs, regulation, and other factors? Electricity suppliers may not undertake such decisions for a variety of reasons, not least of which are the distortions caused by regulation. Nevertheless, concern for cost-minimization should underlie studies of both technological and economic questions affecting the electricity industry. Our results, while tentative, point to the importance of advanced technologies and augmented electricity trade among states for minimizing incremental generation costs. These findings complement other studies of the industry's economic structure in suggesting that the industry's future is unlikely to resemble its past.

Our research program was sponsored in part by grants to Resources for the Future from the Edison Electric Institute (EEI) and the Electric Power Research Institute (EPRI), though both organizations have distanced themselves from the findings in this book. We are particularly grateful to Oliver Gildersleeve at EPRI for his thoughtful, substantive comments on an earlier draft of the study. We also have received comments from many others at EPRI and at individual utility companies.

We are grateful to many colleagues at RFF and academic institutions for their contributions to our work. We have received useful comments on earlier drafts from John F. Ahearne, Douglas R. Bohi, Joel Darmstadter, Karen L. Palmer, Paul R. Portney, and several anonymous reviewers. The model could not have been solved without the software provided by Roy Marsden at the University of Arizona. In addition, we have benefited from countless hours of assistance by Sarah Bales, Fay Kidd, Adam Klinger, Russell Lamb, and Kay Murphy. The able editing by Sam Allen and design work by Brigitte Coulton have produced a volume that is more attractive to look at and easier to understand. As usual, none of the individuals or institutions named above is responsible for the views expressed in the book or for any errors remaining; that responsibility is entirely ours.

Hadi Dowlatabadi
Associate Professor
Department of Engineering and Public Policy
Carnegie Mellon University

Michael A. Toman
Senior Fellow
Energy and Natural Resources Division
Resources for the Future

January 1991

1
INTRODUCTION

Electricity generation in the United States has experienced dramatic changes over the past two decades. Historically the industry faced stable demand growth, fuel prices, and regulation, coupled with ever-larger generating units. These factors contributed to falling real electricity production costs through the 1960s. Starting in the 1970s, however, environmental constraints became more stringent, and fuel prices and demand growth became more unstable. Cost economies resulting from increasing plant sizes or greater thermodynamic efficiency became elusive.[1] Market conditions also became more unpredictable with the introduction of new competitive forces and changes in economic regulation.

Even under the simplest circumstances, choices of technology in electricity generation depend on numerous tradeoffs between capital and fuel costs, between plant refurbishment and new construction, between conventional technologies and new designs, between expansion of capacity and wholesale power purchases, and so forth. The overall impact of the factors noted in the introductory paragraph

---

[1] Traditional design sought economies through larger size, reflecting engineering facts such as the increase in fluid volume relative to boiler surface area as plant size grew. However, this approach became stymied by limits to the properties of materials used for plant construction. Thus, while larger steam plants were cheaper to construct they also were less reliable and costlier to operate. The alternative of pushing thermodynamic limits through "supercritical" technology also proved disappointing, due to plant failures that resulted from impurities deposited in boiler tubes. Both large-size and supercritical plants also have suffered greatly from thermal cycling when attempting to adjust plant output to variations in demand. The result has been an absence of scale returns beyond 600 MW and a move away from supercritical technology. For further discussion, see Joskow and Schmalensee (1985) and Joskow and Rose (1985); see also chapter 3.

has been to confront electric utilities with significant new uncertainties. In the generation sector, uncertainties about demand and economic regulation have clouded the determination of aggregate investment. Uncertainties in fuel prices, unit construction costs, and environmental requirements have further complicated the choice of technologies to be used for new capacity or plant refurbishment. Other important sources of uncertainty are the recent growth of nontraditional power suppliers—for example, industrial plants that cogenerate electrical energy as part of the manufacturing process—and shifts in economic regulation that impose greater market risks on utilities.[2]

This study is concerned with identifying efficient technology choices under different economic conditions and environmental constraints. The methodology used in this study identifies technology choices that minimize the present value of total economic costs for supplying electricity (including construction, fuel, and operating costs) over a planning horizon, subject to environmental constraints and the requirement that demand be served. An important attribute of the methodology is a capability for detailed stochastic sensitivity analysis which helps to identify those factors that most strongly influence our conclusions.

Given the many uncertainties surrounding decisions about investment in generation technology, the methodology and findings of this study are of interest for several reasons. A better understanding of how minimum-cost technology choices might vary with different conditions is important for utility planning and for the determination of research and development programs. For example, to what extent should new investment be directed toward advanced new generation technologies? Are investments in these technologies a good choice under a wide variety of conditions, or do they require special sets of circumstances to be attractive? Are there specific technologies (conventional or advanced) that appear to be particularly strong?

The subjects of the study also are relevant to understanding the impacts of environmental regulation on the generation sector. The identification of technologies that satisfy emission constraints at minimum cost is a central element of our analytical framework. The framework also gauges the marginal cost of electricity supply under different emission constraints. These issues are at the heart of debates over

---

[2] For a review of these changes, see Joskow and Schmalensee (1983), Toman and Darmstadter (1988), and Joskow (1989). Still other important sources of uncertainty are the adequacy of existing transmission capacity, the impediments to expanding capacity, and shifts in regulatory rules for transmission pricing and access. These issues, however, are beyond the scope of the present study.

changes in environmental legislation and the introduction of "environmental costing" procedures to weigh externalities in various regulatory proceedings.

The effects of changes in economic regulation of electricity supply are not a central part of this study. Nevertheless, our methodology can help to clarify some of these issues. For example, if our findings heavily favored the choice of conventional technology over smaller-scale advanced designs, then recent arguments for relaxed regulation of generation markets to promote competitive entry and risk-taking by independent suppliers would be at least somewhat attenuated. Our analysis also provides some information on the advantages of expanded interstate power exchanges.

The following section of this chapter further describes the scope of our analysis. Before proceeding, however, we will note several caveats that apply to the study as a whole. Our focus on economic cost-minimizing technology choices is intended to be neither predictive nor prescriptive. We recognize that costs perceived by utilities may differ from true opportunity costs, thus driving a wedge between the cost-minimizing decisions of utilities and the outcomes indicated by our analysis,[3] and that important equity issues may compete with economic efficiency as social objectives.

Moreover, our interest here is in identifying minimum-cost technology choices in an engineering-economic sense; thus we do not consider factors like the effects of changing economic regulation and shifts in industry structure (such as entry by nontraditional power suppliers). Finally, while the model could be applied to the planning problems of an individual utility, our interest is in broader results about technology choice by aggregates of generating companies at the state or regional level. This more aggregative focus clearly imposes numerous simplifications of reality. In addition, in this study we consider technology choices only for two regions of the United States, and we do not include the full range of possible technology choices.[4] Different conclusions could arise for other regions or with alternative specifications of the technology menu.

These caveats imply that our specific modeling outputs must be viewed as tentative only. Nevertheless, our findings are sufficiently provocative as to indicate at least the need for additional study of

---

[3] This gap is shrinking, however, with the influx of competitive forces in the generation industry and with changes in economic regulations that permit market forces to influence prices.

[4] Our focus on only a subset of states and technologies is intended to keep the study manageable. Chapter 3 explains how the technology set used in the study was constructed.

conflicting influences on the generation technology frontier. These findings point in particular to the desirability of various advanced-design generation technologies for investment in new capacity, and the potential role of expanded interstate trade in lowering electricity supply costs. The results also illustrate the general usefulness of the model for addressing technology choice issues.

Scope of the Analysis

Given our focus on economic cost-minimization over the life of generation investments, several issues must be addressed in the design of a research strategy. Because some of the generation technologies that are candidates for future investment do not yet have significant track records of actual use, and because future market conditions are likely to differ so much from those of the past, standard statistical approaches based on observed data are not applicable. In addition, since decisions about technology choice in electricity generation inherently involve investment in durable assets over time, the analysis of these decisions must be intertemporal. Actual costs incurred over time under different technology choices depend in a complex fashion on the time profile of demand being served, the relative utilization of different plants, fuel costs, and other factors.[5] These influences cannot be captured in simple summary statistics on relative plant costs.

There are numerous "conventional" (tested and in-use) technologies and "advanced" technologies (those being developed or tested in demonstration projects) that could be included in the analysis. Developing a workable analytical framework thus requires judgment about which technologies to include. Since barriers to long-term transactions in bulk power markets may be lowered in the future, the choice of purchasing capacity from other regions must be considered. We also require assumptions that allow us to represent the aggregate of utilities' decisions within states or regions, and we need strategies for representing uncertainties about technology performance and market conditions. All of these facets of the methodology are discussed in the following subsections; see also chapters 2, 3, and 4.

The Basic Framework

We apply a multiperiod, multiregion linear programming model that identifies cost-minimizing plans for investment in new capacity and

---

[5] See Turvey (1968) and Turvey and Anderson (1977).

4

use of available capacity, given an existing capital stock and demand level, plus assumptions about other factors. The model allows us to examine tradeoffs among different generation technologies under a variety of assumptions about market conditions, environmental standards, and other influences.

Utility planning and dispatch models have long been used to study investment decisions over time. Increasingly sophisticated applications of programming and process models have been used for individual utility planning, industry impact analysis, and national policy assessment for more than a quarter of a century. Models have been developed to include regional disaggregation of supply, intertemporal investment planning, market interactions between demand and supply, environmental constraints, and the impacts of economic regulation.[6]

Our model synthesizes and extends many elements found in other frameworks. We combine investment planning over time for multiple regions of the United States with a menu of technology choices for new investment that includes both conventional and advanced designs. The model also includes an endogenous, albeit highly simplified, representation of interstate power exchanges. The model constrains decisions about both new investment and plant operation to satisfy specified sulfur oxide emission limits, which can be met by retrofits on existing coal plants as well as by new investment in lower-emission technologies.

Perhaps the most significant methodological development embodied in our model is a very detailed stochastic sensitivity analysis of investment decisions and supply costs. The technique involves first specifying subjective probability distributions for numerical input assumptions about technology performance and market conditions. The model is then run for a random sample of input assumptions drawn from these distributions. The corresponding probability distributions for the model outputs provide a richer picture of future technology choice than would be possible by focusing on only a few scenarios. Further insights are gained from the use of nonparametric statistical tests to more rigorously identify which parameter variations most strongly affect the model results.

---

[6] For an early example, see Massé and Gilbert (1964). A fairly comprehensive model was developed by Baughman, Joskow, and Kamat (1979), who also provide numerous citations of other examples. Cushey and Rubin (1988) review the application of models for gauging the costs of acid rain control in the United States. Many other models are in use by government agencies, universities, industry trade associations, and consulting firms.

Time Horizon and Technology Options

In this study we assume a planning period ranging from 1985 to 2010. This time horizon allows us to include both commercially available technologies and technologies that are at the pilot or demonstration stage of development today. By including the latter we broaden the relevance of the study to encompass technology development issues as well as capacity planning decisions. Extension of the analysis beyond 2010 would involve even greater uncertainties associated with more speculative technology options.

Of the many advanced technologies being developed, we have selected three options which seem to have significant technological and economic promise. These options are atmospheric fluidized bed coal combustion (AFBC), gas turbine combined-cycle (GTCC), and integrated gasification combined-cycle (IGCC)—the latter a combination of GTCC with a coal gasifier. The conventional alternative for new investment is pulverized coal (PC) steam plants with wet flue gas desulfurization (FGD) equipment to meet existing environmental standards. FGD retrofits for existing coal plants also are included in the model.

In addition, we allow for the possibility of power imports into a state from other states when this might be cost-effective. The cost-effectiveness of imports depends on the delivered cost of imported power as compared with the cost of in-state generation. To assign a delivered cost to imported power, we assume a rule for apportioning the cost reduction between buyer and seller.[7] The results obtained by including this cost specification illustrate what could occur with expanded interstate power trade, but they do not indicate the possible impacts of other pricing rules or of constraints on transmission availability. To provide a point of comparison, we also include results obtained by assuming that only present contracted volumes of power trading occur.

Areas Modeled

Application of the modeling framework developed here to individual utility planning or regional forecasting would require detailed information on demand, generation capacity, and transmission availability

---

[7] Rules of this kind currently are used in many short-term "economy" or "coordination" transactions. In contrast, the transactions being modeled here are for long-term supply. For exporting states, the cost-minimization in the model is subject to the requirement of meeting both native load and export demand.

for individual power control areas.[8] Since utility planning and forecasting are not the purposes of this study, and since the requisite data are not publicly available in any case, we have not pursued this approach. Another option would be to simply invent hypothetical regions (by concocting data on demand and existing capacity), but we would then have no assurance that the results would be even broadly representative of the actual economic frontier for electricity generation technologies.

Our approach in this study involves a compromise between these extremes. The basic "decision unit" in the model for economic cost-minimization is taken to be the state; we thus make the simplistic assumption of perfect intrastate coordination among generating companies. We also allow for the existence of multistate power pools. Projections of electricity demand and capacity in the model start with 1980 data on state demand patterns and capacity, the most recent year for which reliable state-level supply and demand information is available.[9]

To facilitate the analysis of power trading and to strengthen our conclusions generally, we apply the model to two adjacent regions that have sharp differences in (1) the age and technology of their existing plants, (2) the susceptibility of their existing plants to tightened environmental regulations, (3) demand growth, (4) capacity margins, and (5) degree of intraregional coordination. The first region is the five-state area consisting of Indiana, Ohio, Kentucky, Michigan, and West Virginia. Economic cost-minimization calculations are performed separately for these states, though the possibility of power trading among them (and between the regions) is included in the model. The second region includes Delaware, New Jersey, Maryland, and Pennsylvania. Since these states form a tight power pool in practice, this region is treated as a single decision unit in the model.

We emphasize again that the numerical modeling results presented in the study are illustrative but are not intended to be either forecasts or recommendations. To further underscore this point, and thereby to defuse concerns of utility companies or regulators about the nature of our findings, we do not use the actual state names in presenting our results. Instead, we refer to the states in the first group as E1 through E5, where "E" refers to the east central part of the United States and

---

[8] Power control areas are regions within which all power plants are dispatched by the same central authority; plants in a power control area may belong to a single utility or several utilities.

[9] The data are adjusted to account for demand growth and capital turnover since 1980.

the number designates a specific state; we refer to the states in the second group as M1 through M4, where "M" refers to the mid-Atlantic part of the United States and, again, the number designates a specific state.

Sources of Inputs

Running the model requires input assumptions about future growth in total demand, as well as about changes in the shape of its time distribution (for example, changes in the ratio of peak use to average demand). It also requires input assumptions about costs and technical performance of new capacity; changes in fuel prices; the evolution of environmental constraints; and other factors. For the sensitivity analysis, these assumptions must be expressed as probability distributions. Wherever possible, we have built these distributions around forecasts or assessments drawn from government or industry reports. Nevertheless, the location and shape of the input distributions inherently involve a significant dose of subjective judgment.

One advantage of our methodology is its flexibility with regard to input assumptions. If the results of the sensitivity analysis are counterintuitive, or if there are other reasons to question the nature of the assumptions made in running the model, other assumptions can easily be substituted. We return to this point in evaluating our own results in chapter 5.

The Plan of This Book

The balance of this book is organized into six chapters. Chapter 2 describes our analytical framework. Chapter 3 discusses different generation technology options for future capacity investment, and explains the choice of specific technologies included in the study. Chapter 4 contains an explanation of our input parameter distributions, and chapter 5 presents our modeling results. Chapter 6 considers the effects on these results of changing the method by which environmental controls on electricity producers are implemented, thereby illustrating the applicability of the model to policy issues. Chapter 7 briefly recapitulates the main findings of the study and presents some directions for further research.

2

ANALYTICAL FRAMEWORK

The main elements of the analytical framework used in this study are the optimization model for calculating cost-minimizing plant investment and operation plans, and the sensitivity analysis procedure. Other facets of the framework are the organization and preprocessing of data related to existing generation capacity and coal use, and the representation within the model of the cost to importers of bulk power as an incremental source of supply. A schematic representation of this model is given in figure 2-1.

The Optimization Model

The optimization model we use is known as the Planning and Dispatch for Reduced Emissions (PADRE) framework. This model was originally developed at Carnegie Mellon University (CMU) as part of a larger study of how environmental constraints would affect electricity generation.[1] In our application, the model determines the cost-minimizing use of existing power plants and, as these are retired and as demand grows, investment in incremental capacity to meet demand over time. These cost-minimizing decisions can be computed simultaneously for several geographical areas, with cost-reducing power trading included as a possibility.

---

[1] A complete technical exposition of the PADRE model structure is contained in Dowlatabadi, Edahl, and coauthors (1986a, 1986b, 1986c). PADRE was developed as the central component in a suite of models. It was derived from the Multi-Period Multi-State (MPMS) model, which integrated the many modules in the Advanced Utility Simulation Model (AUSM). PADRE has undergone further refinement and testing at Resources for the Future.

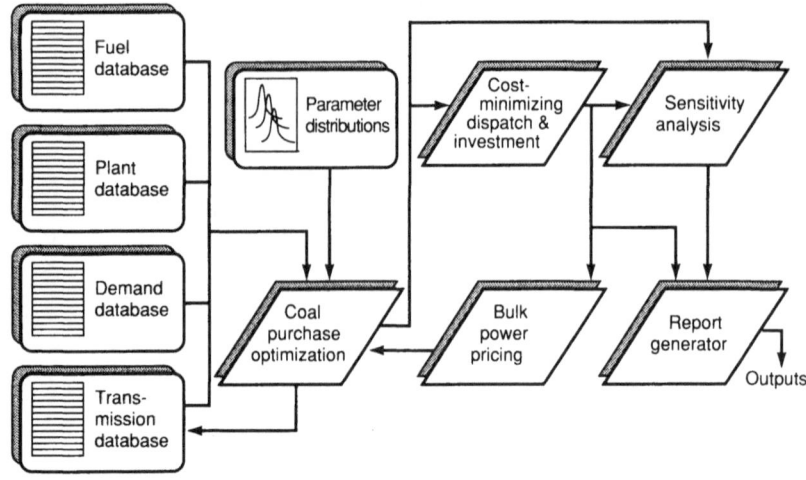

Figure 2-1. A schematic representation of the modeling framework

The cost-minimization problem solved by PADRE can be summarized by the following list of assumptions and constraints, choice variables (decisions and outputs), and the objective of the optimization model.[2]

Assumptions/constraints

- specifications of annual demand *profiles* for the region or regions under study, including changes in these profiles (average growth rates, changes in peak-to-average demand ratios, impacts of industrial cogeneration, and others) over time;
- costs and characteristics of fuels consumed in each region;
- costs and characteristics of generation technologies, including specifications of plant life;[3]
- costs, efficiency, and other characteristics of wet flue gas desulfurization (FGD) pollution control technology;
- transmission costs and a characterization of delivered costs for bulk power imports; and
- specifications of state or regional environmental constraints.

---

[2] The structure of the model is based on the discussion in Turvey and Anderson (1977), chapter 13.

[3] Dowlatabadi (1984) develops another model in which plant retirement or life extension is endogenous.

Choice variables: decisions/outputs

- timing, type, location, and size of generation capacity expansions and FGD retrofits;
- plant dispatch order and quantity of fuels consumed;
- marginal cost of electricity generation by region, period, and demand level; and
- volume of new interstate energy transfers.

Objective

- minimize the present value of total supply costs, including capital, fuel, and operating and maintenance (O&M) costs.

Within the set of existing utility plants, PADRE distinguishes three classes of conventional steam coal plants that are subject to differing environmental constraints; oil-fired steam plants; conventional gas turbines; nuclear plants; and hydro/"other." PADRE also considers up to four conventional or advanced technologies for new investment, as well as FGD retrofits of existing plants. A diverse set of fuels is considered for each region, including three types of coal (distinguished by sulfur content and location) and two types of oil, as well as natural gas and nuclear energy.

Cost-minimization in PADRE is carried out using linear programming (LP). While this approach ignores important nonlinearities, the objective function is easily formulated in this setting and LP formulations involve relatively low computational burdens.[4] A drawback of LP solutions is their "knife-edge" character: small changes in input parameters can induce discontinuous changes in model outputs, raising questions about the robustness of conclusions. Our sensitivity analysis helps to address this concern.

Even with an LP formulation, the complete multiperiod, multiregion cost-minimization problem would remain unmanageable if it could not be divided into smaller parts capable of being solved individually.[5] Spatial decomposition—solving for different states or regions separately—would further simplify the problem but would force us to ignore interactions among areas through expanded interstate power flows. To avoid this, we decompose the problem tempo-

---

[4] The computational size of the model is further reduced by aggregating together some existing power plants in the input database, and by limiting the number of fuel types (see the discussion of coal data below).

[5] There are several standard ways to decompose LP problems (Dantzig and Wolfe, 1960; Lasdon, 1966, 1970; Wismer, 1971), but these methods offer little improvement over "sparse-matrix" simplex algorithms. A more radical approach is needed for our problem.

rally by recognizing first that aggregate demand and thus capacity have grown monotonically in the past, while emission constraints consistently have become more stringent. Given the eminently reasonable assumption that these trends will continue, the model solution for the final period considered will have the largest capacity and the most investment in clean technologies or retrofits. PADRE exploits the observation by decomposing the entire problem into a series of single-period optimization problems solved backward in time. Once the solution for the final period is calculated, the solution for any earlier period simply involves an earlier commitment to the investments in place at the end of the time horizon.[6]

Complications arise with this approach when one technology is a refinement or extension of another. Specifically, in our application the model treats construction of gas turbine combined-cycle (GTCC) and integrated gasification combined-cycle (IGCC) plants as unrelated events. This causes an overestimation of IGCC costs, since in practice IGCC plants can be built in stages—starting as GTCC, followed by the addition of a coal gasifier—with lower overall capital cost.[7] Thus our modeling results are biased against the use of IGCC technology. Given this fact, it is all the more striking that IGCC turns out to be a favored technology in many model runs, as described in chapter 5.

Stochastic Sensitivity Analysis

The first step in the sensitivity analysis involves constructing subjective probability distributions for the model inputs. The probability density functions we assume for each input parameter, and the interrelationships among these distributions, are described in chapter 4.

[6] For further discussion of the methodology, see Dowlatabadi, Edahl, and coauthors (1986a, 1986b, 1986c). Note that this process is not the same as dynamic programming, which characterizes the optimal decision for any state. Here, in contrast, we are computing "path solutions" that depend on the starting point.

[7] Conversion of a GTCC plant to an IGCC plant reduces GTCC capacity in the model, contrary to the assumption of monotonic capacity growth underlying the model's solution algorithm. To partially get around this problem in our solutions we set the upper bound on GTCC capacity in any period t equal to the sum of GTCC and IGCC capacity in period $t+1$ (recall that the model solves backward in time). This allows GTCC investment to be undertaken in recognition of the option for conversion to IGCC. Unfortunately, we also double-count the cost of the combined-cycle investment in this methodology: once at time $t+s$ when the IGCC plant is installed, and again at date t when the GTCC capacity is built. The alternative of running the model twice, once to calculate GTCC capacity and a second time to calculate IGCC capacity given the GTCC plant available for conversion, was computationally infeasible.

Given these subjective probabilities, the next step is to construct the sample of input vectors. Once again we face a tradeoff between exhaustive study and exhaustion of resources. For example, one strategy—a so-called factorial design sampling scheme—would allow each input parameter to assume a specified number of values drawn from its distribution, with the model then being run for all possible combinations of parameter values. In this design, if there are 10 input parameters each assuming three values, a total of 3^{10}—or 59,049—model runs would be needed. Clearly this scheme is practical only for models having a very small number of parameters and corresponding realizations.

A more useful strategy is to ensure that sampling is done from the whole range of possible outcomes for each parameter, even if all possible combinations are not considered. We do this with the so-called Latin Hypercube Sampling (LHS) scheme. Briefly, we use the LHS procedure in the following way.[8] Given the number of model inputs and outputs, as well as the desired power of the nonparametric statistical tests to be used for relating inputs and outputs (see below), we can determine the number of input-output observations that are needed to attain the desired significance levels in the tests. This number, N, is the number of times the model is exercised and the number of times each input parameter is sampled. Given this number, we first divide each input parameter's probability density function into N equiprobable slices. The mean value of each parameter over that slice is then used as the sampled value for the input in the model. The sampling process itself involves the random choice of parameter values/slices without replacement. The input samples cover the full span of each parameter's probability density function.[9]

Application of the model to the sample of input vectors generates corresponding probability distributions for model outputs. The means and variances of these distributions give one indication of the most likely outcomes. However, because the precise numerical outputs convey limited information, given the knife-edge character of the LP solution, we also use two nonparametric tests to examine which inputs have the greatest impact on the outputs. The two approaches we use are known as Kendall's τ and Spearman's rank-

[8] For further discussion of this and other sampling approaches, see Iman and Helton (1985).

[9] These steps must be modified if colinearities are assumed among the input parameters. If that is the case, the input data must be transformed into orthogonal elements before sampling; the parameters used in the simulations then are recovered by reversing the transformation. For details, see Dowlatabadi (1984), chapter 3.

order correlation coefficient ρ_s.[10] In both approaches the numerical distributions of outputs are replaced by distributions defined over the ordinal rankings of these numbers (the same also is done with the input data). The correlation of rankings between inputs and outputs is then examined. If a steady increase in the ranking of an input parameter is found to be correlated with a steady increase (decrease) in the ranking of an output variable, we conclude that there is likely to be a strong positive (negative) relationship between them in practice. In this case the result is sensitive to the input assumption. If instead we find only a weak correlation of input and output rankings, we conclude that the output is robust to changes in that input.[11]

Data Preparation

In addition to (probabilistic) input assumptions for future conditions, the model uses historical databases on existing generating units, demand, and fuel types and costs. In this section we briefly discuss these baseline information sets.[12]

Capacity

Table 2-1 summarizes information from the database on existing plants in 1980 for the two areas, regions E and M, defined in chapter 1. Power plants indicated in the database as being under construction are brought on-line according to their expected completion dates (which also are recorded in the database).

Table 2-1 reveals some interesting differences between the two regions. The most significant difference is in the proportions of coal-fired capacity—more than 80 percent in region E, but less than 30

[10] For more discussion, see Downie and Heath (1965) and Lehman (1975).

[11] Kendall's test is the more nonparametric of the two approaches, using only the relative ordering of input and output ranks, whereas Spearman's statistic uses the numerical ranking values. Kendall's procedure is more robust but less informative. In our analysis we find results with Kendall's test to be robust; this makes us confident of the results obtained with Spearman's statistic, which are the results reported in chapter 5.

[12] The model also contains a database used to estimate costs for new transmission; these inputs are described in chapter 3. This database could be broadened to include information on the existing transmission grid, but such detailed information is not readily available.

Table 2-1. Existing Capacity in Regions E and M (1980)

State	Technology	Capacity (MW)	Heat rate (Btu/kWh)	Availability Maximum (%)	Availability Average (%)	O&M costs[a] (mills/kWh)
E1						
	Oil	1,133	10,073	91.5	87.6	8.37
	Gas turbine	0				
	Nuclear	0				
	Coal-SIP	12,770	10,493	82.0	62.4	2.61
	Coal-NSPS	949	10,771	82.0	63.0	3.30
	Coal-FGD	845	10,550	82.0	62.2	2.51
	Other	0				
E2						
	Oil	299	10,989	91.4	87.1	11.7
	Gas turbine	0				
	Nuclear	0				
	Coal-SIP	10,218	10,364	82.0	60.6	2.91
	Coal-NSPS	0				
	Coal-FGD	1,388	10,646	78.9	60.2	3.96
	Other	740	NA	90.0	45.4	2.80
E3						
	Oil	2,692	10,569	89.0	80.0	1.71
	Gas turbine	475	12,318	89.0	80.0	9.96
	Nuclear	2,827	NA	78.0	58.0	5.43
	Coal-SIP	9,937	10,783	82.0	60.5	3.12
	Coal-NSPS	791	9,719	82.0	57.0	3.30
	Coal-FGD	0				
	Other	2,372	NA	96.3	33.1	6.68
E4						
	Oil	2,220	10,616	91.0	86.1	15.6
	Gas turbine	0				
	Nuclear	906	NA	78.0	58.0	6.90
	Coal-SIP	20,751	10,361	82.0	61.4	3.26
	Coal-NSPS	500	9,700	82.0	57.0	3.30
	Coal-FGD	750	11,153	82.0	57.0	5.10
	Other	0				
E5						
	Oil	0				
	Gas turbine	0				
	Nuclear	0				
	Coal-SIP	11,938	10,046	82.0	66.0	1.83
	Coal-NSPS	1,300	10,771	82.0	63.0	3.30
	Coal-FGD	1,224	10,381	82.0	57.0	3.10
	Other	170	NA	90.0	74.8	NA

(continued)

15

Table 2-1. (*continued*)

State	Technology	Capacity (MW)	Heat rate (Btu/kWh)	Availability Maximum (%)	Availability Average (%)	O&M costs[a] (mills/kWh)
M1						
	Oil	657	11,031	89.0	80.0	1.8
	Gas turbine	0				
	Nuclear	0				
	Coal-SIP	589	10,864	85.0	76.0	3.07
	Coal-NSPS	400	10,800	82.0	57.0	4.0
	Coal-FGD	120	10,643	86.0	70.0	4.0
	Other	0				
M2						
	Oil	3,037	12,136	88.0	77.0	3.96
	Gas turbine	0				
	Nuclear	1,635	NA	78.0	58.0	5.36
	Coal-SIP	2,768	10,451	84.0	69.0	2.85
	Coal-NSPS	0				
	Coal-FGD	0				
	Other	260	NA	90.0	55.8	8.17
M3						
	Oil	5,159	10,988	91.0	85.0	10.73
	Gas turbine	1,566	10,448	89.0	80.0	4.46
	Nuclear	1,699	NA	78.0	58.0	133.66[b]
	Coal-SIP	1,835	10,908	82.0	57.0	4.59
	Coal-NSPS	0				
	Coal-FGD	0				
	Other[c]	330	NA	NA	NA	NA
M4						
	Oil	7,175	10,930	69.0	57.0	4.95
	Gas turbine	0				
	Nuclear	2,896	NA	78.0	58.0	5.66
	Coal-SIP	14,614	10,272	82.0	69.0	2.76
	Coal-NSPS	0				
	Coal-FGD	3,286	10,904	82.0	60.0	6.04
	Other	1,850	NA	9.0	4.5	19.27

Note: SIP refers to coal plants complying with State Implementation Plan Environmental Standards; NSPS refers to plants meeting Environmental Protection Agency New Source Performance Standards; FGD refers to plants having flue gas desulfurization equipment. ''Other'' technologies include: hydro, wind, solar, and waste-combustion, except as noted below. NA = not applicable.

[a] O&M costs are in 1980 dollars, and were gathered from Energy Information Administration, *Historical Plant Cost and Annual Production Expenses for Selected Electric Plants, 1984* (1986) and *Historical Plant Cost and Annual Production Expenses for Selected Electric Plants, 1986* (1988).

[b] This figure reflects a nuclear plant with a very low capacity factor.

[c] Pumped storage.

Source: Carnegie Mellon University power plant database.

percent in region M. Thus we would expect tightened sulfur emission regulations to have a greater impact in region E.[13]

Demand

Baseline demands for the nine states studied in our model are represented by using annual load duration curves (LDCs).[14] A typical theoretical load duration curve is shown in figure 2-2. Each point (τ, δ) on the curve represents the number of hours τ in the year that the level of demand in the state was equal to or greater than δ.

To solve the LP problem in PADRE, we must use a discrete approximation of the continuous LDC shown in figure 2-2. In this study we use a four-block demand representation, as shown in figure 2-3. The four blocks are referred to as peak, cycling, shoulder, and baseload demand.[15] The assumed durations of the blocks are also shown in figure 2-3. These specific durations were chosen to achieve reasonably faithful representations of the continuous LDC in figure 2-2 and of typical utility plant use patterns.[16] The height of each block reflects average demand levels over the block's duration.

---

[13] For completeness, we note that the data contain several simplifying assumptions. All coal-fired units are assumed to have a 75 percent plant availability factor (Electric Power Research Institute, 1986b), though there is evidence that supercritical unit availability averages only about 65 percent (Joskow and Schmalensee, 1985). This assumption understates the need for new capacity and overstates the need for FGD retrofits. A constant average gross efficiency of coal plants over time also is assumed, even though efficiency drops with age while emission rates rise. This assumption understates the need for new capacity and the level of emissions from existing plants. In addition, it overstates the ability of plant operators to switch fuels: with higher emissions than posited, flexibility in fuel choice is more limited. This effect is compounded by the fact that in practice fewer than half of the 120 largest SO_2 emitters can switch coals (Energy Ventures Analysis, Inc., 1985). Finally, system load factors for existing plants in each state and region are biased upward because diversity of demand across regions is not fully taken into account (see the appendix to this chapter); thus a bias toward too much new capacity is built into the model.

[14] The appendix to this chapter contains further information on issues related to demand representation.

[15] Each block corresponds to a different type of power plant typically used to satisfy demands. For example, peak demand is most cheaply satisfied by using plants having a low capital cost, given its brief duration. Conversely, plants for satisfying base demand should have low operating cost, given that they are run more or less continuously; higher capital costs for these plants can be averaged over a large output volume.

[16] The greater the number of demand blocks, the better the representation of cost-minimizing investment. However, the size of PADRE also grows (linearly) with the number of demand blocks, so a balance needs to be struck. Experiments with five-block and six-block LDCs did not show a significant improvement in the fineness of the model resolution over a four-block LDC with a judicious specification of block durations.

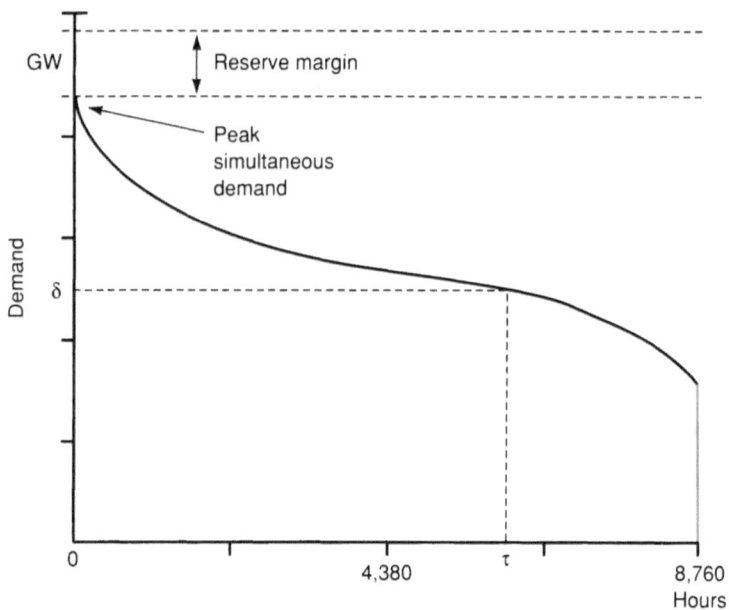

Figure 2-2. Continuous load duration curve (LDC)

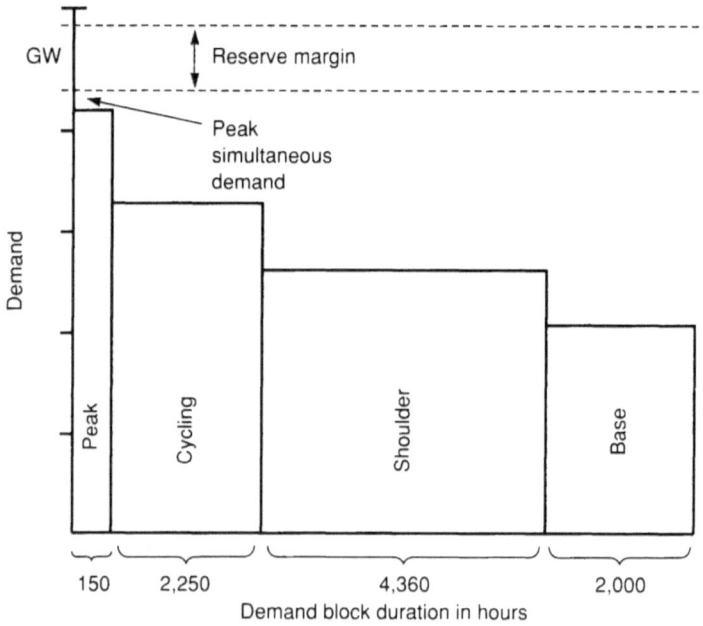

Figure 2-3. Discrete LDC demand blocks

Table 2-2. Structure of Load Duration Curves in the Model

	Percentage of observed peak in demand blocks			
State	Peak	Cycling	Shoulder	Base
E1	92	74	64	50
E2	92	76	64	50
E3	92	82	72	53
E4	96	86	76	60
E5	95	87	76	61
M1	89	71	62	47
M2	89	72	62	48
M3	89	71	62	47
M4	90	76	66	52

Note: The heights of the peak demand blocks in the model are less than the observed peaks because of the discrete nature of the model LDCs; their peak block heights are the average demand levels over the 150 hours of highest consumption in each state.

Source: Calculated by the authors, using raw data from the U.S. Department of Energy.

To parametrize the state LDCs in the model, we begin with raw 1980 figures on hourly generation for Power Supply Areas (PSAs) and construct annual LDCs for the these areas.[17] We then transform these PSA load duration curves into state-level LDCs by first computing county-level LDCs based on each county's population share in its PSA, then aggregating over counties to the state level. Finally, the resulting state-level LDCs are rescaled to reflect actual electricity sales in 1985, the starting year of the model analysis.

Table 2-2 summarizes the structure of the LDCs resulting from this exercise. In all cases the level of peak demand specified in the discrete LDC is less than the observed demand peak because of the averaging process used to define the blocks. The figures in table 2-2 indicate some diversity in the structure of the LDCs. In particular, all of the states in region E have a higher peak, as a percentage of observed peak demand, than the states in region M. Given the nature of the averaging process in creating the LDCs, this implies that the states in region M face steeper actual peaks in their LDCs than the states in region E.

We emphasize that these baseline LDCs represent demands for utility electricity supplies. They take into account whatever levels of

[17] These data were supplied by the U.S. Department of Energy to Carnegie Mellon University and have not been published. Unfortunately, they are the most recent information available for constructing spatially disaggregated LDCs. It should be noted that these data could be used to construct seasonal LDCs (for example, summer and winter), which could be used to derive more detailed descriptions of cost-minimizing investment. However, this task also requires seasonal plant performance data, which are not available, so we focus on annual results.

industrial cogeneration, load reduction, and load management were in place in 1985—programs we assume will continue until the 2010 terminal date used in this study. Of course, much has transpired regarding load management, conservation, and cogeneration since 1985, and there is ample reason to expect further changes in the future. These factors are reflected in our assumptions about future load changes, described in chapter 4.

We also note one other point that relates to the existence of contracts for power transfers. Contracts in place call for region E to export currently more than 3 GW per year, and more than 1 GW per year in the mid-1990s. At present region M is a net contractual importer of more than 1 GW per year, with potential contract volumes of up to 0.5 GW per year in the mid-1990s.[18] These existing contractual transfers are built into the baseline assumptions of the model by subtracting imports from the bottom of a purchasing state's LDC, and adding exports to the bottom of a selling state's LDC. Given existing conditions of capacity and demand, our prior expectation about the model results is that region E would be an exporter of power to region M in many scenarios.

Coal Data

Some preprocessing of the coal data is required since PADRE can accommodate only three coal types per state, out of the more than one hundred types in the database. Moreover, the types of coal likely to be used in practice—and their prices—will vary with the severity of environmental constraints. To reduce the number of options to be considered, before using PADRE, we must run a second LP model designed to select cost-minimizing coal purchases given different power supplies and environmental constraints. This model, which applies to 31 states in the eastern half of the United States, compares delivered costs for coals from more than 40 seams with three different levels of preparation and six different modes of transport.[19] The three coal types in each state that have the highest indicated consumption then are used in the PADRE calculations.

---

[18] These figures are drawn from the Department of Energy (1988). We emphasize that the figures represent only existing contracts, not incremental future transfers from new or renegotiated contracts. As noted, PADRE endogenously calculates the cost-minimizing incremental transfers, given a rule for determining delivered costs. The structure of the rule we assume is described in the text below.

[19] We also assume different coal price differentials related to sulfur content for the various environmental constraints imposed in the analysis; see chapter 4.

Cost of Power Imports

Another preparatory step involves specifying the delivered cost to importers of interstate power transfers. To do this we first run the model under two extreme regimes.[20] In the first we assume no new trade among states in region E or between regions E and M, while continuing to treat region M as a tight pool.[21] This set of assumptions ensures independent capacity dispatch among the individual states in region E and in the region M pool, with no new trade among any of these entities.

In the second regime we run the model to minimize joint costs for both regions, using assumptions for line-loss levels and estimates of marginal cost for incremental transmission.[22] This is equivalent to assuming joint central dispatch and planning of all capacity in both regions. In this regime any trade that can lower joint costs will be undertaken. While trade in the joint-dispatch regime often leads to higher marginal supply costs in power-exporting states compared with costs under the other regime, the total cost of generation for the combined regions invariably is lower.

In practice, of course, prospects for expanded interstate power transfers lie somewhere between these two extremes. To define a middle ground, we first identify all the states that experience reduced costs from new interstate power inflows in the joint-dispatch regime, as compared with their experiences under the no-new-trade regime (for region M states, the focus is on inflows to the region as a whole from region E states). For each such state, the difference in total cost between the two regimes is an upper bound on cost savings that might be achieved from out-of-state power purchases in a situation having less than total regional coordination.

We then assume arbitrarily that half these gains can be captured by exporting states in our middle-ground regime. To do this, we reset transmission charges equal to marginal cost plus half of this trading rent for all states found to be exporters in the joint-dispatch regime. States that were importers in that regime correspondingly are given the option of purchasing power from the exporters at a delivered cost equal to the exporting state's marginal generation cost plus the trans-

---

[20] Recall that the sensitivity analysis involves repeated runs of the model for a random sample of input combinations; the procedure described here is performed for each input combination.

[21] In this regime we do perpetuate existing contractual arrangements for power trading that are in place.

[22] The cost and line-loss estimates we use are summarized in chapter 3.

mission charge, in lieu of using their own capacity. Note that this procedure does not alter the relative cost savings offered by different exporters over in-state supply, but it does reduce the overall attractiveness of imports relative to the joint-dispatch case.

The use of the joint-dispatch scenario as the reference point for computing trading benefits assumes perfect coordination and thus implies greater efficiency than utilities actually can achieve autonomously or in more loosely coordinated arrangements. Thus our formulation overstates the potential trading gains. In addition, our modeling of trading allows greater flows than are possible with the present network. We simply set transmission costs in the model to cover long-run marginal costs and assume that necessary expansion of capacity will be undertaken.[23] We thus are ignoring potential obstacles to expansion of transmission capacity, such as siting and health controversies. These assumptions also lead to an overstatement of gains from interstate power trading.[24] For this reason, in presenting our findings in chapter 5 we include the results without any new power trading. We have already noted that our assumption that half the potential trading gain (as we have defined it) is captured by exporting states is arbitrary; an extension of this study would make the size and division of trading gains the subject of a more detailed sensitivity analysis.

This concludes our description of the analytical framework. In the next chapter we turn to a discussion of the technology options included in the model.

Appendix

Additional Discussion of Demand Representation

Two issues regarding demand representation are discussed here. The first is the use of load duration curves (LDCs) versus "skyline" representations of electricity demand; the second is demand diversity across regions.

Specification of Demand in the Model

All models of capacity planning and dispatch employ some means of demand representation. Short-term models designed for optimization of operations

---

[23] For an illustration of how generation and transmission capacity can be modeled jointly, see Dowlatabadi (1984), chapter 3.

[24] On the other hand, we understate the gains from trade by not addressing demand diversity in our demand assumptions (see the appendix to this chapter).

have a time horizon of a few days and make use of hourly demand data in the form of forecasted daily demands. In longer-term models aggregated annual demands are used. In the latter models the diurnal variation in demand over the whole year is characterized by discrete demand levels or blocks, each lasting for a specified period. Two common representations have been used to this end: skylines, which involve aggregations of demand based on time of day, day of week, and seasonal indices; and LDCs, which are based on the level of demand.

In figure 2-4 a typical daily load curve is depicted. The diurnal cycle in demand follows the daily pattern of human activity. The skyline representation in figure 2-5 has average demand levels for weekdays in summer and winter, for nights in summer and winter, and for weekends. The width of each of the demand blocks in figure 2-5 represents the total time duration for that demand category. The height of each demand block is the average of all the demands realized in that period.

The skyline demand representation is preferable if the timing of power plant operation is of interest—for example, if one's interest is in plant use on winter days between 12 noon and 8 p.m. However, because of the approximate uniformity of demand levels in each block, skylines are inappropriate representations of demand variability and provide a poor representation of demand for use in capacity planning models. Hence if capacity expansion rather than operational issues are of interest, LDCs should be used. LDCs allow the supply sources used for each demand block to be better matched

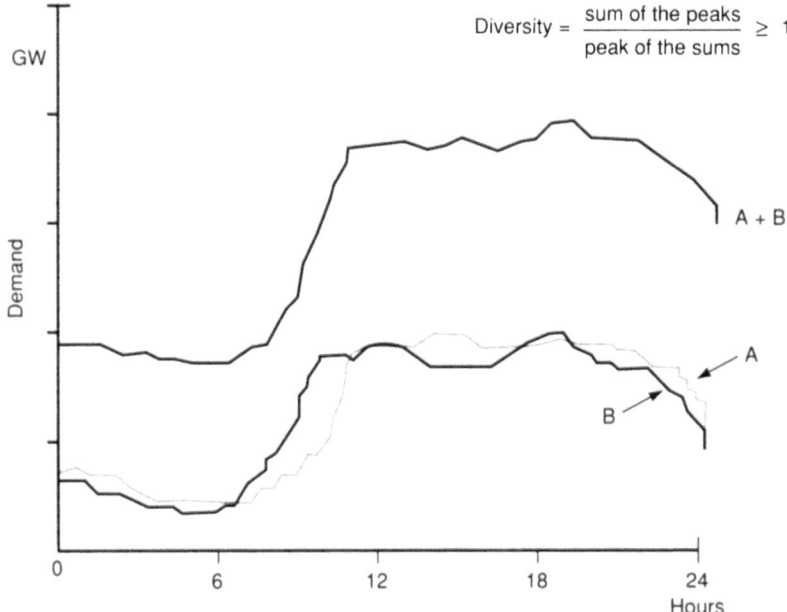

Figure 2-4. Daily load variations

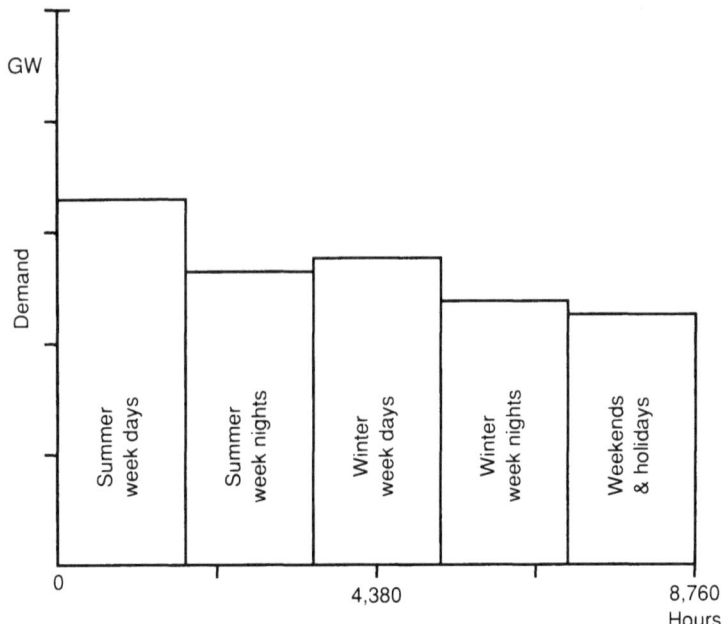

Figure 2-5. "Skyline" demand blocks

with the economic and operational characteristics of various technologies. Since the primary focus of our study is on capacity investment, we use the LDC representation.

Demand Diversity Factors

Suppose two hypothetical adjoining regions, A and B, have the load curves shown in figure 2-4. The demand profiles of both regions exhibit the same diurnal cycle, but the peak demand in these regions occurs at different times. This means that the plant built to meet the peak in region A may also contribute to meeting the peak in region B. The metric for this temporal diversity of demand is equal to the ratio of the sum of the peak demands to the peak of the sum of the demands.

The gains possible from such cooperative capacity planning and dispatch are well known and are exploited in tight power pools. Although the diversity factor is also of great importance in the benefits achieved through electricity trade, it is not reflected in this modeling exercise. This unfortunate compromise is forced on us by the absence of accurate means for calculating diversity factors.[25]

---

[25] As noted in the text, the state demand figures have been calculated using a renormalization of Department of Energy figures for Power Supply Areas. The boundaries of PSAs usually cross over many states, rendering the renormalized state demand figures more temporally correlated than is the case in practice.

24

Demand diversity also figures in the calculation of capacity reserve margins. The usual modeling strategy is to constrain the capacity of power plants to exceed expected demand by a specified reserve margin (RM), as shown in figure 2-2. This margin is set so that given past availability of plants, the chance of a capacity shortage at peak demand is below a specified limit.[26] For a multistate model, the diversity of demand allows for a lower absolute RM, due to the low frequency of concurrent peaks in different states. Ideally, treatment of the RM would need to take into account the reliability of the intraregional transmission links. Unfortunately, given the information available, we are unable to make an accurate determination of the reliability of intraregional links.

[26] In the United States, on average about 17 percent of the installed capacity is not available for electricity generation. However, historically the uncertainty in projecting future demand levels has meant that reserve margins have been set at higher values— the actual value used by most utilities has been close to 22 percent. In Great Britain, where the highest loads are experienced during winter, the loss of load criterion through inadequate capacity has been set at no more than four winter peaks per century (approximately 28 percent).

3

SELECTION OF TECHNOLOGY OPTIONS

Our choice of technologies to include in the model reflects economic and environmental considerations, along with practical constraints imposed by the model structure. In this chapter we first review general changes in the economics of construction costs of different plants, as well as the advantages of modular designs. We then introduce environmental and other considerations to complete our determination of the technology menu.

Changing Power Plant Economics

Two basic doctrines have historically guided the design and construction of power plants. The first is that there are significant physical economies of scale in building plants: other things being equal, cost grows less than proportionately to plant output with an increase in the size of generating units. This follows from the fact that cost is related to the surface area of the boiler, steam generator, and turbine, while output is related to plant volume, which grows faster than surface area with an increase in plant size.[1] In addition, thermodynamic losses are related to surface area (a ''sink'' of energy input), while steam production is related to plant volume. Thus, other things again being equal, larger plants should have lower electricity production costs due to greater thermal efficiency. In practice, this doctrine

---

[1] More precisely, this doctrine states that plant costs are proportional to capacity raised to the two-thirds power.

has led to production of boiler-turbine-generator sets having a capacity as large as 1,500 MW.

The second doctrine that has shaped the design and construction of power plants is related to the operating temperature and pressure of steam generators and turbines. The physics of electricity generation using steam dictates that higher thermodynamic efficiency is associated with higher steam temperature and pressure: higher temperature and pressure entering the turbine imply higher energy content and thus greater electricity output per volume of steam input. This doctrine has led in particular to the development and building of supercritical boiler-turbine-generator sets that generate very high steam temperature and pressure (up to 1,200° F and 4,500 psi).

Recent experience has been at variance with these doctrines, however. The real construction cost of coal-fired power plants rose more than 100 percent between 1965 and 1982. Part of this cost increase reflects additional equipment and technical complexity required to meet environmental protection constraints. Nevertheless, part of the increase also reflects lower productivity at the construction site, which seems to be at least partly a result of larger plant sizes (see Joskow and Rose, 1985). Larger plant sizes also have been associated with lower plant availability and the resulting need to rely upon expensive backup power. Both lower construction productivity and reduced plant availability have worked to offset physical scale economies associated with larger plant size.

Supercritical technology yields a modest increase in the engineering efficiency of electricity generation—roughly 2 to 3 percent. However, supercritical plants also have been dogged by problems of availability (see Joskow and Rose, 1985, and Joskow, 1987). This factor and the higher cost of building supercritical plants have offset their greater thermal efficiency.

Experience with larger and supercritical plants suggests that economies of scale probably are exhausted at a relatively low unit scale of roughly 600 MW (see Joskow, 1987). Larger plants also have the disadvantage of longer lead times (see Rothwell, 1986). Longer lead times not only expose the electricity producer to higher capital costs from interest expenses, but also expose the producer to greater risks associated with demand growth, fuel prices, and regulatory treatment.

Advanced technologies embodying "modular" designs avoid many of these problems associated with conventional steam coal plants. Modular units are smaller in size, 150–250 MW, and are largely assembled in factories before shipment to the plant site. The units of smaller size forgo some physical economies of scale, but they gain economies of mass production for the equipment manufacturer.

Modular plants also compare favorably to conventional designs with respect to thermodynamic efficiency when one takes into account a range of possible plant capacity utilization factors. This phenomenon is illustrated in figure 3-1. Panel A illustrates the variation in thermodynamic efficiency of a single 600 MW power plant with changes in capacity utilization. Panel B depicts the variation in thermodynamic efficiency of a plant composed of four 150 MW units with changes in total capacity utilization.

As the maximum efficiency of a plant is positively related to size, the maximum efficiency of a single 150 MW unit does not match the efficiency of a 600 MW plant. However, when loads of less than 600 MW are to be met, the modular plant can operate with some portion of its units at maximum capacity and thus achieve greater total efficiency than the partly loaded 600 MW plant. It follows that if there is concern about the ability of a plant to meet uncertain demand growth or demand fluctuations experienced as cycling loads, the overall performance of the modular plant is likely to equal or exceed the performance of the conventional single-unit plant. Similarly, when the conventional plant ages and is shifted from base load to cycling load because of higher operating costs, its advantage in thermodynamic efficiency over a modular plant will be eroded.

When one also takes into account the advantages of modular designs in mitigating financial risks—additional units can be added as needed, reducing interest costs and the danger of overinvestment—it is clear that in this study we must consider modular plants as an alternative to conventional steam coal plants. Our specific choices of modular technologies for inclusion in the study are dictated by the relative performance and cost of different options in meeting environmental constraints, among other factors.

Design of the Technology Menu

The current structure of the model allows us to identify up to four technologies as candidates for investment in new capacity. We can also incorporate a technology for retrofit of pollution control equipment to existing coal plants.[2] Because there are a number of new

---

[2] Another option available for reducing emissions from coal combustion is the use of different methods for cleaning the coal before combustion. This possibility is taken into account in the preprocessing of coal price and quality inputs to the model, discussed in chapter 2. Because the focus of this study is on choice of generation technology rather than on coal markets per se, we do not go into detail here on the application of coal-cleaning technologies.

Panel A

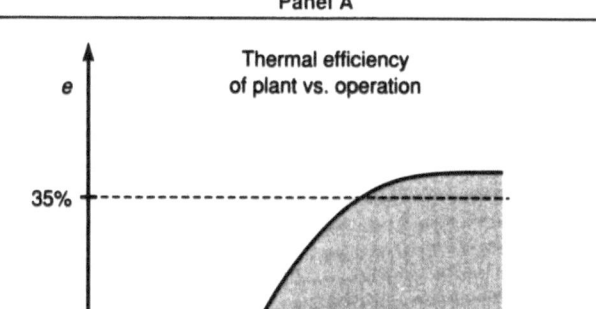

One × 600 MW generation unit

Panel B

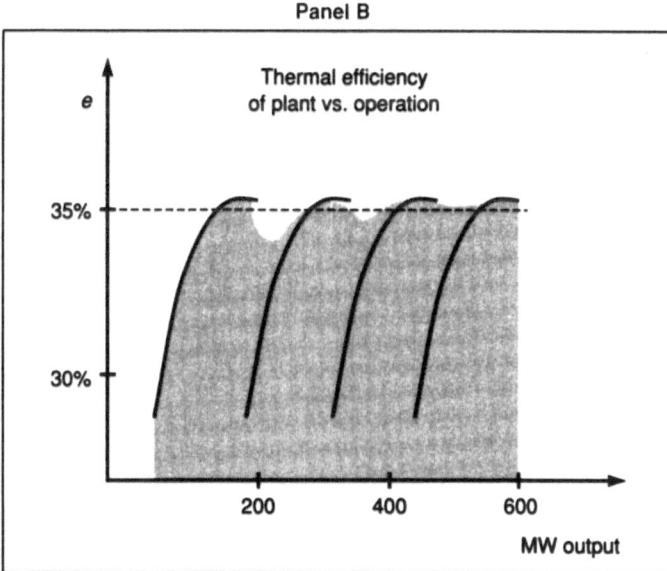

Four × 150 MW generation units

Figure 3-1. Operational benefits of modular design

capacity or retrofit technologies that we could consider, the structure of the model itself imposes nontrivial constraints on the range of possible model outcomes. Those technology choices that we do include in the model have been selected on the basis of properties of the technologies as reported in the Electric Power Research Institute's (EPRI) *Technical Assessment Guide* (*TAG*) and a Department of Energy (DOE) report, along with other sources of information (see Electric Power Research Institute, 1986b, and Department of Energy, 1987).

Because our focus in this study is on fossil-fuel technologies for new capacity and retrofit investments, we do not consider new investments in technologies based on renewable energy inputs or nuclear power plants. Certain renewable energy technologies in certain locations are competitive at present. However, more practical and more widely applicable renewable energy technologies appear to remain some years into the future. In this study we restrict attention to technology options having a stronger medium-term competitiveness, both in the nine-state area that is our focus and nationally.

Nuclear plants in principle have offered the tradeoff of higher capital costs for lower fuel and operating costs. In practice, however, nuclear investments have been stymied for a decade by a crisis of confidence on the part of the public and investors that reflects cost overruns, poorer-than-expected plant performance, safety concerns, and an unresolved waste disposal dilemma. Interest in nuclear power could be revived in the future, particularly with development of new plant designs and with the growth of concern over greenhouse gas emissions. Nevertheless, we focus here on fossil-fuel technologies, leaving a possible exploration of the nuclear option to future work.

Our attention in this study is on technologies likely to be used by utilities or by independent suppliers under long-term contract to utilities. Industrial or municipal cogeneration technologies constitute another supply option that could be explicitly included in the model as an alternative to new utility capacity construction. This option is of particular interest in light of the growing coordination between utilities and cogenerators, which is leading to greater possible dispatch of cogenerated supplies. Another option that could be considered is conservation and load-management services to obtain greater energy services from a given volume of electricity supply.[3]

Unfortunately, our efforts to include cogeneration as a supply option have been stymied by difficulties in obtaining data on these technologies in a form compatible with the information requirements of the model. The cost and performance attributes of cogeneration

---

[3] See Cicchetti and Hogan (1988) and references therein for a review of these concepts.

Table 3-1. Alternative Environmental Retrofit Technologies

	Type	Sulfur removed (percentage)	Cost (mills/kWh)
Combustion	Lime injection multistage burner (LIMB)	50	5–8
	Slagging coal combustor	50–90	1–2
Post-combustion	Flue gas desulfurization (FGD)	>90	9–11

Source: Department of Energy (1987a).

technologies vary widely, and we have been unable to obtain information on the full range of possible technologies or the relative prevalence of their use. Under these circumstances, we have been unable to construct a meaningful characterization of a "representative" cogeneration supply option to include in the technology menu. The problems faced in including conservation and load management are even more daunting. Our solution is to indirectly include all of these options in the study through changes in the demands utilities face, as if these programs were mandatorily undertaken. The procedures used for adjusting state-level LDCs to reflect cogeneration, conservation, and load management are discussed in chapter 4. Further research could include these options in the technology menu if the requisite technical information could be obtained.

Given our primary focus on fossil-fueled utility plants, we next consider various options for environmental retrofits to existing coal plants. Table 3-1 lists three possibilities. The lime injection multistage burner (LIMB) and slagging combustor are experimental technologies that remove sulfur during coal combustion.[4] Flue gas desulfurization (FGD) systems are end-of-pipe technologies that capture sulfur from plant exhaust after combustion.[5]

---

[4] The LIMB process involves the introduction of a sorbent which interacts with sulfur during combustion to form suspended particles that can be removed from the flue gas stream by using already proven methods for fly ash collection (baghouses or electrostatic precipitators, for example). In the slagging coal technology, the radial and axial flows of air and fuel are carefully controlled for more efficient combustion; this method also lowers NO_x production.

[5] Current FGD systems use a wet sorbent, which is used up in the sulfur-removal process, thus forming significant quantities of solid waste. More advanced FGD systems, currently being demonstrated in Europe and Japan, promise to have lower capital cost and a regenerable sorbent. With this technology the large volumes of waste generated by conventional FGDs would be avoided and operating costs likely would be lower. In addition, advanced FGDs may be excellent scavengers of NO_x. However, because the availability of these technologies within our time frame is uncertain, we have not included them in the study.

The information in table 3-1 reveals a set of tradeoffs between cost and effectiveness of sulfur removal. Of the options considered FGD removes the most sulfur, but it is the most expensive (costing up to $250/kW to install, and using up to 8 percent of net power generated at the plant); it also generates substantial volumes of solid waste. We have used FGD as the choice of retrofit technology because it was best able to bring about compliance with the rigid performance standards of the Clean Air Act before the 1990 revisions of the act. The 1990 revisions provide new cost-effective options for compliance; these issues are discussed in chapter 6.

Because the size and other characteristics of an optimally designed retrofit depend on boiler size and the fuel type used, we cannot assume a generic retrofit unit. Instead, we adapt Rubin's (1983) model of new FGD installation in order to describe the optimal retrofits for each existing plant in the model's database. In adapting Rubin's model we arbitrarily inflate his cost levels by 30 percent to account for the higher cost of retrofitting versus new construction. We also assume that the energy cost of operating the FGD unit is the marginal cost of electricity generation as endogenously determined within the model. The units generally cost more than $200/kW to install and add from 9 to 11 mills per kWh to the operating cost of the plant. Part of this higher operating cost reflects the value of the roughly 8 percent of net plant energy output consumed by the FGD unit.

Since there is also the option of interstate power purchases for meeting demand that is in the model, we must specify parameter values for transmission costs and line losses for new power transfers. Ignoring uncertainties, we set these parameters using a set of simplifying assumptions. Transmission costs are calculated as a linear function of the distance between the ''generation centroids'' of importing and exporting states. The centroids reflect the distributions of existing generation plants across each state in terms of location and power output.[6] The distances between the centroids in the nine-state area are shown in table 3-2.

The calculation of the cost coefficient for these distances is illustrated in table 3-3. In this calculation, transmission voltage is assumed

[6] More precisely, we calculate a generation centroid by assigning a ''mass'' to each plant equal to its distance from the geographical center of its state and its relative share of total state generation output. The point at which all of the masses balance is the generation centroid. (This process can be visualized by thinking of a board in the shape of the state placed on a needle fulcrum, with weights reflecting the masses of the power plants resting on the board; the generation centroid is the point where the board with weights physically balances on the fulcrum.)

Table 3-2. Distances Between Generation Centroids in the Nine-State Region (in miles)

State	E1	E2	E3	E4	E5	M1	M2	M3	M4
E1	—	174	216	190	302	571	510	625	329
E2		—	354	232	241	534	466	608	343
E3			—	176	329	502	458	526	242
E4				—	108	385	327	435	139
E5					—	293	225	368	108
M1						—	69	93	262
M2							—	155	218
M3								—	294
M4									—

Source: Calculations by the authors.

Table 3-3. Specification of Unit Transmission Cost Assumptions

Component	Value
(1) Construction cost for double-circuit 345 KV line	1.1×10^6/mile
(2) Amortization factor (costs spread over 14 years)	.072
(3) Number of hours in average year (allowing for leap year)	8,766
(4) Line capacity, fully loaded	415×10^3 kW
(5) Assumed typical line loading	0.5 (50 percent)
(6) Cost add-on factor for transformers, etc.	1.2 (20 percent increase)
(7) Transmission cost: $[(1) \times (2) \times (6)] \div [(3) \times (4) \times (5)]$	0.05 mills/kWh-mile

Sources: The construction cost estimate in (1) is adapted from Electric Power Research Institute (1986b); the entries in (2), (4), (5), and (6) reflect assumptions by the authors.

to be 345 KV and the lines are assumed to be double-circuit. Transmission losses for new power transfers are assumed to be 3.6 percent per hundred miles of transmission distance.

Before considering possible technologies for new investment capacity, we should briefly mention two other technology options related to existing capacity: plant life extension and repowering. Extension of a plant's life beyond its design age through more intensive equipment maintenance and refurbishment can be an attractive way to forestall capital costs for new construction. In fact, this has been an option of choice for many electric utilities. Unfortunately, the structure of the PADRE model does not permit a detailed analysis of this option. For this study, therefore, we arbitrarily assume that existing plants are operated at their existing power output levels for twenty years beyond their design ages—a life extension that seems to be a defensible approximation of much recent experience.[7]

Repowering basically involves replacing the boiler (and often other equipment, such as fuel-handling facilities) with equipment of more advanced design while retaining much of the other equipment in the plant (outer building, transformers, control systems, turbines). The replacement equipment is likely to reflect the same advanced designs that are found in the technology we consider for new investment. While repowering requires higher capital costs than simple plant life extension, it offers the potential of expanded plant output and significantly lower air-polluting emissions at a cost that is less than the cost of new capacity. Thus repowering is an important option in any full consideration of technology choices.[8]

A meaningful treatment of the repowering option, however, requires a plant-specific analysis to take into account the idiosyncratic features of various existing plants. Such an analysis is beyond the scope of our current model, with its generic technology categories. Exclusion of this option biases upward our results on the size of new investment and FGD retrofits. The effect on our results concerning the composition of investment is unclear, but it could well be minor given the similarity of the technologies that are candidates for investment in new capacity and in repowering.

We turn now to the choice of the four technology options for new investment to be included in the model, noting again our focus on fossil-fueled technologies. One of the four choices must be traditional pulverized coal (PC) in order to provide a reference point for comparison with advanced modular designs. To satisfy environmental constraints, we assume that any new PC plants are equipped with wet FGD units. (The 1990 revisions of the Clean Air Act make possible other technical options for meeting environmental constraints, but we do not consider them here.) Our assumptions about the costs and performance of PC/FGD technology are specified in chapter 4.

Another natural candidate for inclusion is the gas turbine combined-cycle (GTCC) plant. This is a modular technology with quite low capital cost (around $600/kW, in 1985 dollars) and high thermodynamic efficiency. Fuel cost with this gas-fired technology is higher than with coal plants, but there is no sulfur pollution stream. This

---

[7] Dowlatabadi (1984) describes a framework in which decisions about plant life extension and plant retirement are endogenous. Unfortunately, this framework is not generally applicable to the analysis of generation and trade across multiple regions.

[8] For further discussion of repowering, see Department of Energy (1987a, 1987b).

technology clearly is an important option for meeting future peak and cycling demands, and in some circumstances it could even play an important role in meeting base and shoulder loads.

Two slots in the technology menu remain open. In line with our discussion in the first section of this chapter, we seek two other modular design options. Table 3-4 provides information on three such options. Two of these designs are fluidized bed systems that capture sulfur during combustion. These systems also feature highly efficient heat transfer and lower NO_x emissions than conventional pulverized coal units. Of the two designs, pressurized fluidized bed combustion (PFBC) operates at higher pressure than atmospheric fluidized bed combustion (AFBC), thus permitting higher combustion efficiency and temperature in a smaller boiler. Both designs do generate significant amounts of solid waste by-products.

The third design listed in table 3-2 is the integrated gasification combined-cycle (IGCC) system. Unlike the other coal technologies considered here, in an IGCC system the coal is first converted to gas, then the gas is used to fire a combined-cycle turbine. Coal gasification is extremely efficient at sulfur removal and is also very effective at reducing other pollutants (see Wolk and Holt, 1988), though it also generates large volumes of solid waste. Coal gasification is not competitive at current gas prices, but is expected to become competitive as gas prices rise over time. IGCC units also are expected to have high reliability and thermodynamic efficiency.

Although PFBC appears to offer significant economic advantages over AFBC, PFBC development has been plagued with difficulties in

Table 3-4. Alternative Modular Coal Technologies

Type and (TAG code)[a]	Technical status	"Overnight" capital cost (1985 $/kW)[b]	Sulfur removal (percentage)
AFBC (10.2)	Pilot	1,500	85–90
IGCC (14.2)	Demonstration	1,400	>90
PFBC (12.1)	Laboratory	1,500	90

[a] AFBC = atmospheric fluidized bed coal combustion; IGCC = integrated gasification combined-cycle; PFBC = pressurized fluidized bed combustion. TAG codes refer to specific types of technologies listed in the *Technical Assessment Guide* (Electric Power Research Institute, 1986b). The code numbers in this table identify the particular "species" of technologies used in this study.

[b] "Overnight" cost estimates exclude interest expenses for funds used during construction.

Sources: Electric Power Research Institute (1986b); Department of Energy (1987a).

the past.[9] Research on PFBC is continuing.[10] Given the uncertainty in the outcome of PFBC development, we have elected to include AFBC and IGCC systems to round out our technology menu. Inclusion of these two different technologies increases our ability to draw sharp conclusions from the model outputs.

To sum up, the menu for new investment in our application of the model consists of flue gas desulfurization retrofits, conventional pulverized coal with FGD, and three modular technologies: gas turbine combined-cycle, atmospheric fluidized bed coal combustion, and integrated gasification combined-cycle. In the next chapter we discuss the assignment of specific values to input parameters for these technologies and for other components of the model.

[9] The most ambitious PFBC project to date, at Grimethorpe, England, was abandoned in 1988.

[10] Much of this work is being undertaken by the Department of Energy and the American Electric Power Company under DOE's Clean Coal Technology Program.

4

INPUT PARAMETER DISTRIBUTIONS

Our sensitivity analysis starts with an explicit representation of uncertainties about the inputs through the specification of probability distributions over the input parameters. Wherever possible, we try to specify ranges for these distributions that span the forecasts by other organizations, such as the Energy Information Administration (EIA), Data Resources Incorporated (DRI), the Electric Power Research Institute (EPRI), and the North American Electricity Reliability Council (NERC). Where sources for parameter specifications are not otherwise indicated, the choices reflect our own judgment.

It should be noted that the input parameters in this study are for the most part assumed to be independent of one another.[1] In practice, of course, many parameters are interdependent. For example, fuel prices and electricity demand are related, but in focusing on supply decisions rather than overall market behavior, our approach does not exploit knowledge about how demand responds to fuel prices. Fuel costs also influence the unit construction cost of generation plants. Furthermore, much of the technology employed in the generation of electricity is shared among different plant designs. Thus there are numerous colinearities in the construction cost and thermodynamic efficiency of various technologies. The scope of this study does not allow us to address these issues.[2]

---

[1] The exceptions involve fuel prices and environmental constraints, as discussed below.

[2] For discussion of a more sophisticated sensitivity analysis that does address these points, given the requisite data, see Dowlatabadi and Evans (1986). By not taking parameter correlations into account we are, in effect, placing too much weight on unlikely parameter combinations. The result is more diffuse probability distributions for model outputs than would be obtained with more discriminatory input sampling.

Demand

Numerous factors must be considered in specifying how the load duration curves (LDCs) may change over time. Basic growth in demand, load management by utilities, and changes in the level of cogeneration all can play significant roles. The expected future evolution of each of these factors is considered below.

Basic Growth in Electricity Demand

Different sources provide widely varying forecasts of the growth in total demand for electricity over time. For example, the average annual growth rate for total demand (net energy for load) is projected by the North American Electricity Reliability Council (1987) to be about 2.1 percent for the continental United States over the period 1987 to 1996. NERC also presents high and low annual growth rates of 3.5 and 0.9 percent, depending on an unspecified relationship between electricity and "key energy-related variables." NERC's forecasts are based on utilities' own judgments about demand growth and thus already reflect some assessments of load management and nonutility generation. Using a computerized modeling system, the Energy Information Administration projects a national annual growth rate for electricity demand of 2.4 to 2.7 percent over the period 1987 to 2000, depending on oil prices and economic growth.[3]

We build uncertainty about basic demand growth into our sensitivity analysis by assuming that in both region E and region M the average annual growth rate for total demand over the period 1985 to 2010—before accounting for the effects of load management and cogeneration—is a random variable distributed uniformly over a range of 1.5 to 3.0 percent.[4] This range seems wide enough to encompass most of the growth rates in basic demand that are found in different publicly issued projections.[5]

---

[3] See Energy Information Administration (1988). This source does not provide state or regional breakdowns in its forecasts.

[4] To simplify the design of the sensitivity analysis, we apply each of the growth rates sampled from this distribution to all of the states in regions E and M. An extension would allow for more diversity in state-level demand growth by sampling each state's growth rate separately.

[5] Judgments elicited from experts at EPRI would suggest the use of a broader distribution ranging from 0.9 to 3.5 percent per year. Our demand modifiers, discussed below, spread the growth rate projections beyond the 1.5–3.0 percent range.

Strategic Conservation and Industrial Cogeneration

Further uncertainty about demand growth arises from the potential for expanded industrial cogeneration or increased conservation induced by utilities through load management programs. A recent EPRI study suggests possible total peak savings for the United States from cogeneration and conservation in the year 2000—beyond savings already in place in 1983—of 14 to 156 GW, with a median value of 54 GW.[6] These savings compare to a projected peak demand in 2000 of 715 GW without the programs. Thus the savings represent from 2 to 22 percent of projected peak load, with a median of around 8 percent.[7]

The range of growth rates in the NERC report reflects some accounting for expected greater conservation and expansion of industrial cogeneration. To combine uncertainty about basic growth rates with the uncertainty about future load reduction, we first make the simplifying assumption that the basic demand projections already reflect the median load savings given in the EPRI report—that is, 8 percent in the year 2000. We then shift the EPRI report's distribution of savings to this center, so that the 10th percentile is −6 percent and the 90th percentile is +14 percent. The lower end of this translated distribution reflects higher load growth than expected due to less successful conservation programs or slower growth in cogeneration. Similarly, the upper bound reflects more success than expected in limiting load growth.

Our means for combining uncertainty about basic demand growth, conservation, and industrial cogeneration into the sensitivity analysis is best described algebraically. Let γ represent the average annual growth rate in demand (given average success in slowing load growth through cogeneration or conservation), distributed uniformly between 1.5 and 3.0 percent. Let α represent the additional fractional load reduction in the year 2000 from sources described above—that is,

---

[6] See Electric Power Research Institute (1986a). The end points of the range are the 10th and 90th percentiles of the distribution of savings. The figures here are slightly different than those in the EPRI report in that we include nonutility-induced, demand-reducing cogeneration, and we allow for about 5 GW of conservation in 1983.

[7] Slightly higher savings are anticipated in off-peak loads. The figures in the EPRI report are not necessarily inconsistent with the roughly 20 GW of nonutility generation recorded in Edison Electric Institute's 1985 survey (Edison Electric Institute, 1987). That survey includes, for example, roughly 6 GW of gas-burning cogeneration in the southwestern part of the country. We would classify much of the capacity providing this power as using ''utility'' generation technology, given the way we have distinguished utility and industrial generation technologies.

α has 10th and 90th percentiles of -6 percent to $+14$ percent, with a median size of zero. Assume (as in the EPRI report) a smooth growth of conservation and cogeneration over time. Let γ' denote the average annual demand growth that reflects the cumulative impact of load reduction between 1983 and 2000, in that the year 2000 load with growth rate γ' is equal to the load with growth rate γ minus the load reduction given by α.

Then, assuming a time horizon of T years,

$$(1+\gamma')^T = (1-\alpha)(1+\gamma)^T$$

or, solving for γ',

$$\gamma' = (1-\alpha)^{1/T}(1+\gamma) - 1$$

In our sensitivity analysis, we combine values of γ drawn from its distribution with values of α drawn from our renormalization of the EPRI report's distribution for load savings to generate the adjusted growth rates γ'.[8] In practice we use a lognormal representation of α. The effect of this procedure is to extend somewhat the range of possible growth rates, and to reshape the distribution for the demand growth rate to a convolution of a uniform and a lognormal distribution. For example, where T is equal to 17 years (1983–2000),

with $\alpha = +14$ percent and $\gamma = 1.5$ percent, γ' falls to -6 percent;
with $\alpha = -6$ percent and $\gamma = 3.0$ percent, γ' rises to 3.4 percent.

Thus our procedure effectively extends the range of possible demand growths from a low of -6 percent to a high of 3.4 percent.

Load Management

The factors examined above are also likely to change the shapes of the load duration curves. Specifically, many utilities have already launched programs designed to manage the growth in peak demand, as well as total sales. However, there is some justification for caution in evaluating the potential impact of load management programs. One reason for this is the structural shift in the economy from manufacturing to service industries, which tend to exhibit more of a peak in electricity usage during normal business hours. Another reason is the

---

[8] We are assuming that the possible percentage load savings computed nationally in the EPRI report can also be applied to regions E and M.

limited incentive for consumers to participate in load management programs in the absence of pricing that is closer to the short-run marginal cost of electricity.

The EPRI study of load management projects impacts of new measures for peak-clipping and load-shifting in the year 2000 (beyond programs in place in 1983) that range from 0 to 66 GW, or 0 to 9 percent of peak demand, with a median reduction of 17 GW, or 2 percent of peak demand. We have also noted the sharp growth in peak demands in the past two years. Thus we choose to define a probability distribution for load management that includes the possibility of "peakier" LDCs. The lower end of the load management peak-reduction distribution is therefore extended arbitrarily from 0 to −2 percent (that is, there is a possibility of up to 2 percent net growth in the peak).

The possible impacts of load management on LDC shapes are brought into the model in the following way. For any adjusted demand growth rate γ', the peak growth rate is reduced by an amount drawn from a translated lognormal distribution with the characteristics just described. The negative values on the probability distribution functions are used to reflect the possibility of a growth in "peakiness" despite load management programs. In this modification of the load duration curve, total sales are held constant while the areas in the peak and cycling blocks are increased (decreased) in the same amount that the areas under the shoulder and base demand blocks are decreased (increased).[9]

Cost and Performance of New Capacity Investment

Our sensitivity analysis also includes uncertainties about the cost and performance of additions to generating capacity. As explained in chapter 3, the four technologies considered in the model for new capacity are atmospheric fluidized bed combustion (AFBC), gas turbine combined-cycle (GTCC), integrated gasification combined-cycle (IGCC), and pulverized coal with flue gas desulfurization units (PC/FGD). We do not consider costs and performance uncertainties associated with the operation of existing plants or with the retrofitting of

---

[9] This assumes that reductions in peak demand arise from load-shifting rather than peak-clipping, whereas the EPRI study projects that about two-thirds of the reductions will come from the latter programs. However, since the ratio of peak demand volume to total demand volume is small (as can be inferred from figure 2-3), the distortion introduced by this assumption is limited.

existing coal plants with FGD units. Our assumptions regarding existing plants and FGD retrofits are described in chapter 3.

Selected characteristics of these technologies, as specified in EPRI's *Technical Assessment Guide (TAG)*, are outlined in table 4-1 (many specific types of PC, IGCC, GTCC, and AFBC technologies are listed in the *TAG*; the *TAG* code numbers given in table 4-1 and subsequent tables identify the particular "species" of technologies used in this study).[10] We assume a unit size of 200 MW for GTCC, IGCC, and AFBC, and a 500 MW unit size for PC/FGD. To test for the sensitivity of our results to this assumption, we also undertake one separate run of the model (outside the main sensitivity analysis) with a 500 MW unit size for AFBC.

To perform our sensitivity analysis, we need to create probability distribution functions for many of the parameters in table 4-1. Such functions for the availability of units and their heat rates cannot be obtained from published sources. The uncertainty about availabilities is reflected in the model entirely in our assumptions about unit construction costs.[11] The uncertainties in heat rates, shown in table 4-2, are based on our own judgments. We assume asymmetric triangular distributions over the ranges shown, with modal values given by the figures in table 4-1.

Our treatment of uncertainty in unit construction cost makes use of uncertainty bounds presented in the *Technical Assistance Guide* and those calculated from empirical evidence in the chemical industry. The *TAG* gives error bounds for unit construction cost estimates, ranging from ±10 percent for mature technologies like pulverized coal to ±20 percent for technologies at the pilot plant stage, such as atmospheric fluidized bed combustion. However, a study by the Rand Corporation finds that actual costs of new *industrial* (chemical) process plants tend to be consistently underestimated. Furthermore, the degree of underestimation becomes progressively larger as the technology involved becomes less mature (see Merrow, Phillips, and Myers, 1981). To the extent that this conclusion is also applicable to electricity generation processes, the Rand analysis suggests that the

---

[10] See Electric Power Research Institute (1986b). Note that while the total capital cost of PC/FGD is set as an input to the model, the portion of O&M costs related to operating the FGD unit is calculated endogenously, just as are the costs of FGD retrofits on existing plants (see chapter 3).

[11] The *TAG* gives no information for constructing a sensitivity analysis on plant availabilities, so we use the point estimates given in table 4-1. It should be noted that uncertainties about availability and capital cost are dual in that lower expected availability results in higher capital outlays for obtaining sufficient capacity to meet demand, so our uncertainty ranges for capital costs can be viewed as proxies for the uncertainty in plant availability.

Table 4-1. Engineering-Economic Characteristics of Various Generation Technologies

Technology and (TAG code)	Fuel type	Heat rate (Btu/kWh)[a]	O&M cost (mills/kWh)[b]	Availability Maximum (%)	Minimum (%)	Capital cost ($/kW)[b,c]
Atmospheric fluidized bed combustion (10.2)	Coal	10,045	7.8	81.5	81.3	1,508
Gas turbine combined-cycle (44.2)	Gas	8,480	2.1	95.0	90.3	527
Integrated coal gasification combined-cycle (14.2)	Coal	9,775	3.2	94.7	83.2[d]	1,411
Pulverized coal with FGD (5.2)	Coal	9,800	2.0[e]	77.9	71.2	1,106[f]

[a] Represents average annual heat rate.
[b] All costs are in December 1984 dollars.
[c] Unit construction costs for ''overnight construction'' (no interest expenses for funds used during construction are included).
[d] The minimum availability figure is assumed to be close to the equivalent availability figure published in EPRI's *Technical Assistance Guide.* A more precise calculation of this figure using the methodology outlined in Electric Power Research Institute (1979) would yield a figure of 81 percent.
[e] Does not include the operating cost of the FGD unit, which is computed endogenously (see text).
[f] Includes costs of both the PC plant and the FGD unit.

Source: Electric Power Research Institute (1986b).

Table 4-2. Heat Rate Uncertainties for Various Technologies

Technology and (*TAG* code)	Status	Heat rate (Btu/kWh)		
		Low	Modal	High
AFBC (10.2)	Pilot	9,750	10,045	11,000
GTCC (44.2)	Mature	8,230	8,480	10,050
IGCC (14.2)	Demonstration	9,490	9,775	9,980
PC/FGD (5.2)	Mature	9,760	9,800	10,350

Note: The form of the assumed distribution is asymmetric triangular. The modal value of heat rates presented here coincides with figures published in Electric Power Research Institute (1986b). There are no published data on the uncertainty bounds for the heat rates of various generation technologies. Here we have used part-load unit efficiencies to mimic these bounds, the lower bound being that for a 50 percent load and the upper bound for a 100 percent load. These figures are approximations and do not reflect EPRI data on actual plant performance uncertainties.

TAG unit construction cost estimates given in table 4-1 may be too low to accurately represent a realistic modal estimate; and that even if they do serve this purpose, an asymmetric distribution with a longer tail for costs above the modal level may be appropriate.[12]

To address this concern, we assume asymmetric triangular distributions for unit construction costs. The modal estimates for these distribution functions are set equal to the *TAG* mean values. The lower bounds for unit construction costs are also drawn from this source.[13] However, the upper bounds of the distribution are derived from information in the Rand study.[14] The greater asymmetry in these distribution functions for unit construction costs raises the median values above the mode, as shown in table 4-3. The difference between the

[12] To be sure, industrial and electricity generation processes are different; our use of the Rand findings to adjust the bounds on construction cost error is intended to be illustrative only. The larger question here is that there is not only uncertainty about the cost and performance of new generation technologies, but also uncertainty about the uncertainty. Both uncertainties are manifest in disagreements (and divergent investment plans) within the industry and EPRI. As noted previously, one strength of our methodology is that the sensitivity analysis allows us to identify results that are counterintuitive or dependent on a specific set of input conditions. In chapter 5 we discuss the sensitivity of our model findings to our assumptions.

[13] The lower bound is set at 10, 15, and 20 percent below the *TAG* mean for mature, demonstration, and pilot technologies, respectively.

[14] The Rand report gives statistical estimates of the mean ratio of planned to actual costs, and a one-standard-deviation confidence interval around the mean (see Merrow, Phillips, and Myers, 1981, pp. 37–38). We use the reciprocal of the lower bound of the interval as a measure of the maximum degree to which costs exceed their modal values. Since the interval includes only one standard deviation above and one below the mean, our upper bound is conservative. The Rand numbers are presented for differing degrees of technological maturity, ranging up to a ranking of 4 for the most mature technologies. We have translated the Rand ranking to the *TAG* characterization of technologies by setting 4 = mature, 3 = demonstration, and 2 = pilot.

Table 4-3. Unit Construction Cost Uncertainties for Various Technologies

Technology and (*TAG* code)	Unit construction capital cost distribution (1985 $/kW)			
	Minimum	Mode	Median	Maximum
AFBC (10.2)	1,206	1,508	1,669	2,412
GTCC (44.2)	474	527	538	620
IGCC (14.2)	1,199	1,411	1,511	1,987
PC/FGD (5.2)	995	1,100	1,128	1,301

Note: The form of the assumed distribution is asymmetric triangular. The figures here reflect the authors' subjective assessment of the range of unit construction costs for various technologies. The modal values coincide with the figures presented in Electric Power Research Institute (1986b). However, the uncertainty in the cost distributions also has been derived from other findings, in particular Merrow, Phillips, and Myers (1981).

mode and median for the various technologies can be seen to decrease with the maturity of the technology. In order to test the sensitivity of our findings to these risk assumptions, we undertake a separate experiment (outside the main sensitivity analysis) in which the risk classifications for AFBC and IGCC are reversed. The results of this experiment are described in chapter 5.

Three tacit assumptions regarding advanced technologies and pollution issues should also be noted. In specifying the cost of adding AFBC capacity, we have assumed that there are no constraints on sites for disposing of the solid waste generated during combustion. If this assumption is wrong, then the full cost of AFBC investment, with waste disposal included, could rise substantially above what we have assumed, making this technology less competitive. We also have assumed that solid wastes from new coal gasification plants would not have concentrations of carcinogens that would make the wastes subject to strict standards for hazardous-material disposal.[15] Finally, all of our assumptions about investment and fuel costs ignore the possible impacts of regulations to limit nitrous oxide or carbon dioxide emissions.

Fuel Prices and Environmental Regulations

We describe here our input assumptions regarding growth rates in fuel prices, environmental constraints, and the impacts of these constraints on coal prices. The fuels in question include residual fuel oil,

---

[15] See Bombaugh and Rhodes (1988). Our basic assumption is that carcinogenic wastes can be avoided through tighter control over the chemistry of the gasification process, even though the solid waste of the gasifier is likely to have a higher ratio of active to inert materials than bottom-ash or fly-ash from conventional plants.

natural gas, and coals differentiated by sulfur content.[16] Coal prices are assumed to be correlated with environmental constraints.

Growth in Oil and Gas Prices

We have examined fuel price projections from both the Energy Information Administration (EIA) (1988) and Data Research Incorporated (DRI) (1987), but we rely primarily on EIA projections, for several reasons. The DRI and EIA base case projections for oil and gas prices are almost identical. EIA also gives price projections under three well-defined market conditions, thus providing some basis for varying the price assumptions in our sensitivity analysis. EIA fuel price projections for the period 1987–2000 are summarized in table 4-4.[17] The price trajectories for oil and gas appear to be correlated, as one would expect.

In the sensitivity analysis for changes in petroleum prices, we assume for simplicity that possible values for oil and gas price growth rates over the period 1987–1990 are perfectly correlated along a (parabolic) curve passing through the low, base, and high growth-rate pairs given in table 4-4. We make the same assumption for price growth rates for 1990–2000. We then assume triangular distributions for the possible growth rate pairs along these curves, with modal values equal to the base case growth rates in the table.[18] In sampling possible growth rates in the sensitivity analysis, we also assume perfect correlation in the growth-rate pairs between the two subperiods—that is, high growth rates for 1987–1990 are associated with high growth rates for 1990–2000.

Since this study extends to the year 2010 and EIA's projections end in 2000, we develop supplementary petroleum price growth rates for the period 2000–2010. We assume that gas and oil prices grow together over this period at a percentage rate equal to the average growth of oil prices from 1987 to 2000. Thus the sensitivity analysis for petroleum prices over the period 2000–2010 is based on a triangu-

[16] For simplicity we combine distillate and residual oil-fired plants, thus do not consider distillate fuel separately. Nuclear fuel also is included in the model. However, because the decision to operate nuclear plants is essentially not dependent on considerations of their fuel costs (which are low relative to other fuels), only a single point estimate of nuclear fuel costs is used.

[17] We use actual prices for 1985 and 1986 in the execution of the model. Also, we treat oil and gas prices separately from coal prices in the analysis, since the projections suggest little cross-price sensitivity between petroleum prices and coal prices during the period in question.

[18] While the growth rates are assumed to be correlated, their magnitudes differ, so that the relative price *levels* of oil and gas vary across the distributions.

lar distribution for a common growth rate of both gas and oil prices. This assumption is made to allow for some moderation in oil price growth as its price level rises, and to allow for a closer movement of oil and gas prices after a catch-up period of more rapid growth in gas price through the 1990s. We further assume that high price growth before 2000 is correlated with high price growth afterward, as was assumed for the 1987–1990 and 1990–2000 periods. It should be noted that all the price trajectories show smooth growth over each sub-period; our analysis does not incorporate occasional temporary price shocks.

The EIA oil and gas price projections have been criticized for not showing enough sensitivity to market conditions. Some critics have argued that the projections generally are too low (see, for example, Hogan, 1988). In addition, the average annual price growth rates over the period 1987–2000 are somewhat bunched together across the three EIA market cases (particularly for natural gas, whose growth rate as shown in table 4-4 ranges from 4.9 percent to 6.3 percent). Given our strategy for constructing fuel price probability distributions, the result of this bunching is fuel prices that do not show large variations in growth rates from one execution of the model to another.

This narrowness does somewhat impede our ability to assess the consequences of much higher gas prices on new capacity investment. Since growth rates compound over time, in our analysis the absolute difference in the year 2010 between the lowest and highest possible gas prices is substantial—$5.50/million Btu as compared with $8.00/million Btu (in 1985 dollars). Moreover, even with the lower range of gas price growth rates there is a significant increase in the price of gas

Table 4-4. EIA Real Oil and Gas Price Average Per-Annum Growth Rate Projections (in percentages)

Fuel	Base case		
	1987–1990	1990–2000	1987–2000
Natural gas	3.4	6.4	5.6
Residual fuel oil	−0.2	5.1	3.8
	Low world oil price/high growth case		
	1987–1990	1990–2000	1987–2000
Natural gas	2.1	5.8	4.9
Residual fuel oil	−5.6	4.9	2.3
	High world oil price/high growth case		
	1987–1990	1990–2000	1987–2000
Natural gas	3.8	7.1	6.3
Residual fuel oil	3.2	6.6	5.8

Note: All percentages represent changes in delivered costs to electric utilities.
Sources: Energy Information Administration (1988a).

Table 4-5. Alternative SO₂ Emission Constraints for States East of and Adjoining the Mississippi River

Case	Million tons of SO_2 emissions from utility boilers per year					
	1985	1990	1995	2000	2005	2010
1	14.5	14.5	14.5	14.5	14.5	14.5
2	14.5	14.5	12.5	9.5	4.5	4.5
3	14.5	12.5	9.5	4.5	4.5	4.5
4	14.5	12.5	4.5	4.5	2.5	2.5

Source: Authors' assumptions.

relative to coal by 2010—roughly 3.5:1, as compared with rough parity in 1987. We discuss the sensitivity of our results to fuel price assumptions in chapter 5.

Environmental Constraints and Coal Prices

Four alternative assumptions about environmental regulation that are considered in the sensitivity analysis are shown in table 4-5. These environmental standards represent a wide range of possible emission control regulations for sulfur dioxide (SO_2), out to the year 2010. In case 1, emissions are held constant at the level that would have been achieved in 1980 given compliance with State Implementation Plan (SIP) emission standards developed in response to the Clean Air Act.[19] Cases 2, 3, and 4 represent accelerated emission reduction schedules of 10 million tons per year (cases 2 and 3) and 12 million tons per year (case 4).[20] We assume these emission reductions are distributed across power plants so as to minimize the costs of compliance within each region E state and the region M pool. We briefly consider the implications of other methods for allocating compliance efforts in chapter 6.

Given our technology assumptions, utilities are likely to respond to the requirements for reducing emissions through a combination of fuel-switching and FGD use. Thus coal prices and environmental constraints cannot be treated independently. In case 1 we assume for simplicity that there is no (real) growth in any mine-mouth coal price; coal transport prices rise at rates up to 1.5 percent per year, depending on the transport mode.[21] In cases 2, 3, and 4 we reason that an

---

[19] Actual SO_2 emissions in 1980 are estimated to have been between two and three million tons above the levels which would have obtained had compliance with the SIP regulation been achieved.

[20] Cases 2 and 3 are variations of the SO_2 control regulations proposed by President Bush in June 1989.

[21] Transport costs are simulated using another model; see chapter 2.

accelerated schedule for emissions control will lead to a sustained shift toward lower-sulfur coals, resulting in premium prices being paid for these fuels.

We model these premiums as differential growth rates in mine-mouth prices, using figures drawn from an EPRI study of how resource depletion affects coal prices (Electric Power Research Institute, 1984). These differentials are shown in table 4-6. In case 1, there are no changes in the relative price of coals. In cases 2, 3, and 4 we postulate differentials representing a "premium coal" market, starting at successively earlier future dates. For each of these cases, the year the differentials begin corresponds to the year in table 4-6 when a sharp reduction in emissions (toward 4.5 or 2.5 million tons of SO_2 per year) is assumed to be mandated. Once again, the delivered prices of coal in the future are also assumed to reflect growth in transport costs.

Discount Rates

Because the model describes investment planning over multiple periods, it requires an assumption about the discount rate used to compute present values of costs over time. The rate should represent the marginal opportunity cost of capital, and since other cost assumptions are expressed in real terms, the rate should be net of inflation. We set the median value for this rate in the sensitivity analysis at 9 percent. This figure reflects a 12.4 percent generic rate of return on equity published at the start of 1988 (see Federal Energy Regulatory Commission, 1987), and a projected (by DRI) 1988 inflation rate of 3.4 percent. Around this median value, we assume a symmetric truncated normal distribution that ranges from 6 to 12 percent. This range can be interpreted to reflect possible changes in overall conditions in

Table 4-6. Price Growth Differentials for Various Coal Types Under Alternative Emission Reduction Constraints

| Case | Year differentials begin | Sulfur content (lbs of SO_2/mmBtu) | | | |
		< 1.2	1.2–1.8	1.8–3.0	> 3.0
1	—	0.0	0.0	0.0	0.0
2	2000	1.0	0.50	0.0	0.0
3	1995	1.5	0.75	0.0	0.0
4	1990	1.5	1.5	0.5	0.0

Note: All entries represent differential price growth rates for each alternative specification of environmental control (see table 4-5).

Source: Derived from Electric Power Research Institute (1984).

49

the financial market, or in the relative riskiness of investment in generation. A separate experiment with a high discount rate of 18 percent (in conjunction with modal values of all other parameters) also is undertaken to investigate the consequences of higher discount rates on the optimal strategy for meeting electricity demand.

Summary of Input Assumptions for the Sensitivity Analysis

The evolution of demand is assumed to be composed of four elements. The basic growth rates are uniformly distributed between 1.5 and 3.0 percent. These are modified by the load reduction from co-generation and strategic conservation savings, which are taken to have a translated lognormal distribution with a median of zero, a 10th percentile of −6 percent, and a 90th percentile of +14 percent. The load duration curve is assumed to respond to load management. This response occurs through the shifting of peak and cycling loads to shoulder and base periods. The magnitude of the shift is given by a translated lognormal distribution with a median of 2 percent, a 10th percentile of −2 percent, and a 90th percentile of 9 percent.

Four technologies are included in the model for incremental additions to capacity: atmospheric fluidized bed combustion, gas turbine combined-cycle, integrated gasification combined-cycle, and pulverized coal. Table 4-1 gives best-estimate values for engineering-economic characteristics of these technologies; uncertainty about heat rates of new technologies is illustrated in table 4-2; and uncertainty about construction costs of new technology units is shown in table 4-3. These distributions are asymmetrically triangular in form, with longer upper tails to reflect possible cost overruns.

Reference points for oil and gas price growth from 1987 to 2000 are drawn from the Energy Information Administration; for coal, price premiums for sulfur-content and transport-cost escalations are derived from information from the Electric Power Research Institute. Oil and gas prices are treated separately from coal prices in the sensitivity analysis. Possible oil and gas price growth rates for 1987–1990 and 1990–2000 are assumed to be perfectly correlated across their respective ranges (see table 4-4). Possible price growth rates for 2000–2010 are assumed to be the same for oil and gas, with a range equal to possible oil price growth rates over the period 1987–2000. Over each of these ranges, an asymmetric triangular distribution is assumed, with the modal value equal to the base case value from the EIA high and low price-growth cases. Price growth trends are assumed to be perfectly correlated across the three subperiods.

Four alternatives for sulfur emissions control, assumed to be equally probable in the sensitivity analysis, are shown in table 4-5. Changes in environmental controls produce corresponding price growth differentials for coals having different sulfur content, as shown in table 4-6. The real discount rate in the sensitivity analysis is drawn from a symmetric truncated normal distribution with a median value of 9 percent and a range of 6 to 12 percent.

We turn next to the results obtained from our analysis.

5

FINDINGS FROM THE MODEL

We exercise the PADRE model under two sets of assumptions about growth in interstate power trade, as discussed in chapter 2. For one regime we assume no increase in power trading among the region E states and the region M pool beyond existing contractual volumes. For the other regime we assume that individual region E states and the region M pool can freely pursue options for outside supplies that would have lower costs than the costs of indigenous supply options. In the new-trade regime, the delivered cost of imports reflects generation costs, transmission costs, and a supplier premium reflecting half of the total potential cost savings from imports to the buying state. Total potential cost savings in turn reflect the difference between a state's supply cost with no new trade and its cost with full regional coordination.

In this chapter we first present results from the model on technology choice and the size of new capacity investment under the new-trade regime. We then compare findings with and without new interstate trade flows to illustrate the sensitivity of our investment results to trade opportunities. Reverting next to the new-trade regime, we present the conclusions of our sensitivity analysis in order to highlight the importance of various input assumptions. As we have pointed out, the findings presented are intended to be illustrative, and are not forecasts or recommendations about utility investment. In the final section of the chapter we summarize our results and comment further on their strengths and weaknesses.

Technology Choice and Investment Size Under the New-Trade Regime

Our findings on technology choice and investment size with new trade are summarized in a series of probability histograms, presented in figure 5-1, panels A through E. The histograms indicate incremental capacity investments and FGD retrofits if trading is allowed beyond existing contract volumes. The histograms are produced by running the model 60 times for different input parameter vectors, constructed through the stochastic sampling process described in chapter 2. Each histogram represents a discrete probability density function. Thus in figure 5-1, panel A, there is about a 40 percent probability (frequency) of observing gas turbine combined-cycle, (GTCC) investment in the 0–10 GW range, a 33 percent chance of observing GTCC investment in the 10–20 GW range, and so forth.

Probably the most striking feature of our results is the lack of investment in new atmospheric fluidized bed combustion (AFBC) capacity in any of the scenarios (hence there is no histogram for AFBC investment). This result deserves careful explanation, particularly since AFBC is a favored technology for new investment by many in the electricity generation industry. Part of the explanation no doubt lies in the knife-edge character of the linear programming solution in PADRE; AFBC could have been a "nearly cost-minimizing" choice in many scenarios. However, the fact that none of the model runs yields AFBC investment indicates that other factors must be considered in explaining the model outcomes.

At least part of the explanation derives from our input assumptions concerning capital costs and technology performance.[1] AFBC and integrated gasification combined-cycle (IGCC) units are both suited to meeting baseload demands, and they have roughly comparable construction lead-times (three years for IGCC and four years for AFBC). However, the nature of our probabilistic input assumptions is such that unit construction cost for IGCC in the model is lower in most cases than unit construction cost for AFBC. This arises from the assumptions that AFBC has a higher expected capital cost and a greater probability of high capital cost (see table 4-3). Moreover, the assumed thermal efficiency of AFBC units is lower in most cases than that of IGCC (see table 4-2).[2]

---

[1] Demand and fuel price assumptions are less likely to weigh heavily in the choice between AFBC and integrated gasification combined-cycle (IGCC): since both are coal-fueled baseload technologies, we would expect them to be similarly affected by changes in these assumptions. It should be recalled that capacity decisions are calculated backward in time by PADRE. As a consequence, the selection by the model of IGCC over AFBC does not reflect low initial gas prices and the ability to convert GTCC units to IGCC units (in fact, the model results are biased against IGCC because these aspects are ignored).

[2] The lower thermal efficiency of AFBC reflects lower assumed combustion temperatures to reduce emissions.

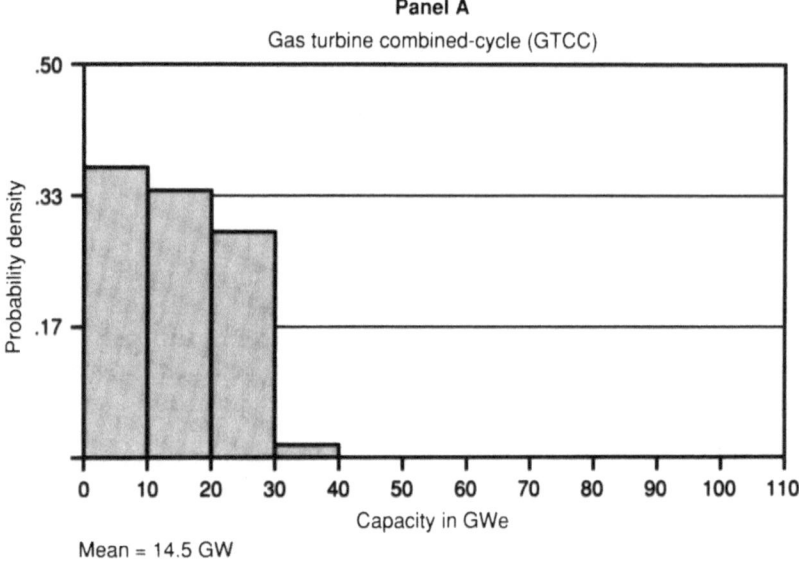

Panel A

Gas turbine combined-cycle (GTCC)

Mean = 14.5 GW

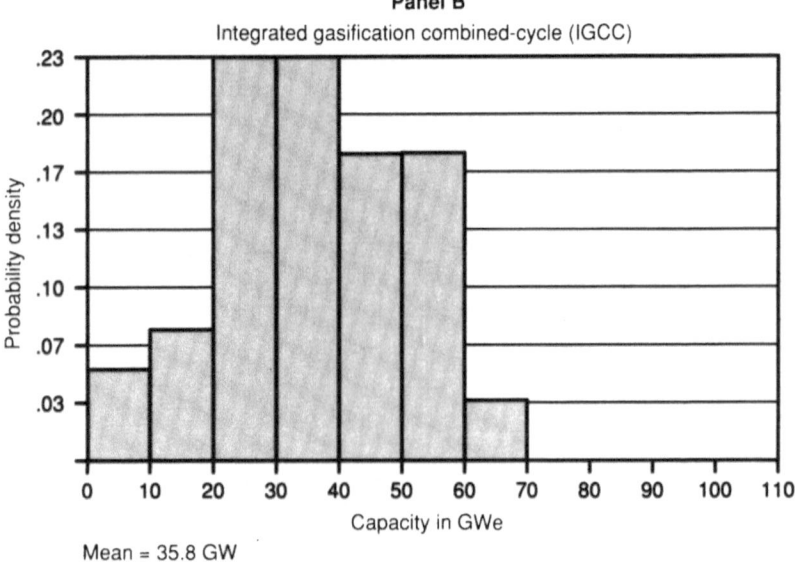

Panel B

Integrated gasification combined-cycle (IGCC)

Mean = 35.8 GW

Figure 5-1. Investment histograms when additional interstate power trading is allowed

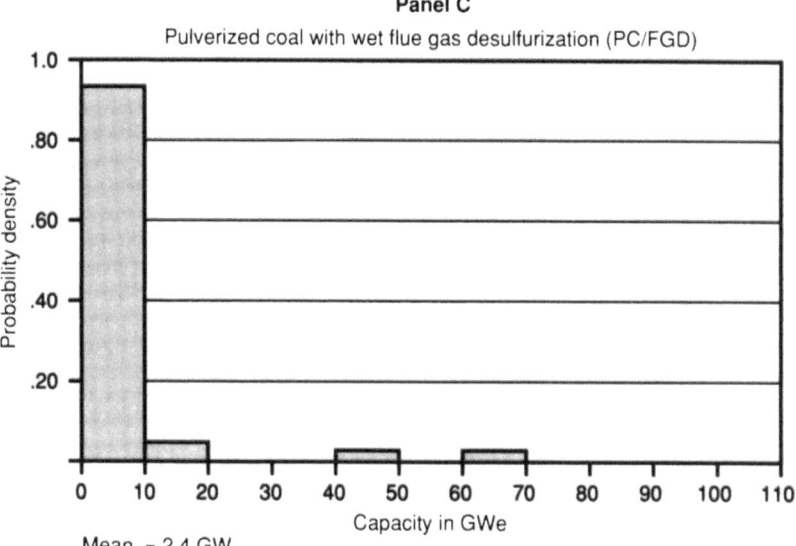

Panel C

Pulverized coal with wet flue gas desulfurization (PC/FGD)

Mean = 2.4 GW

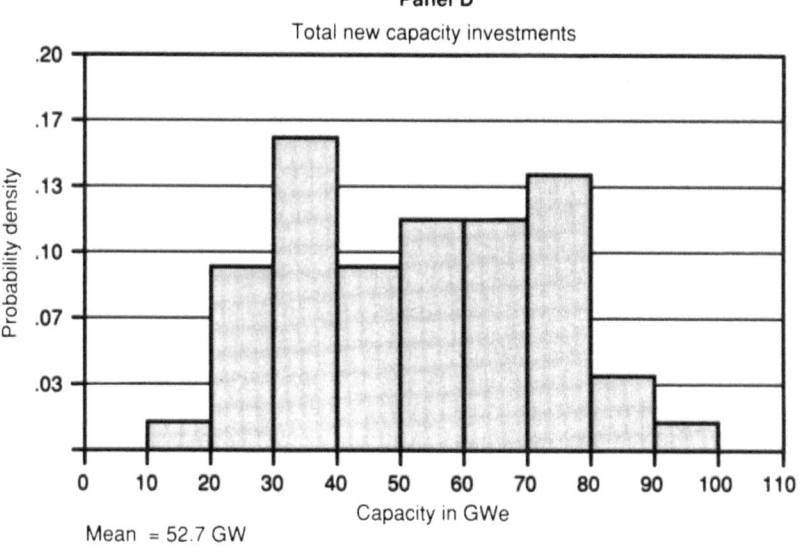

Panel D

Total new capacity investments

Mean = 52.7 GW

Figure 5-1.

(continued)

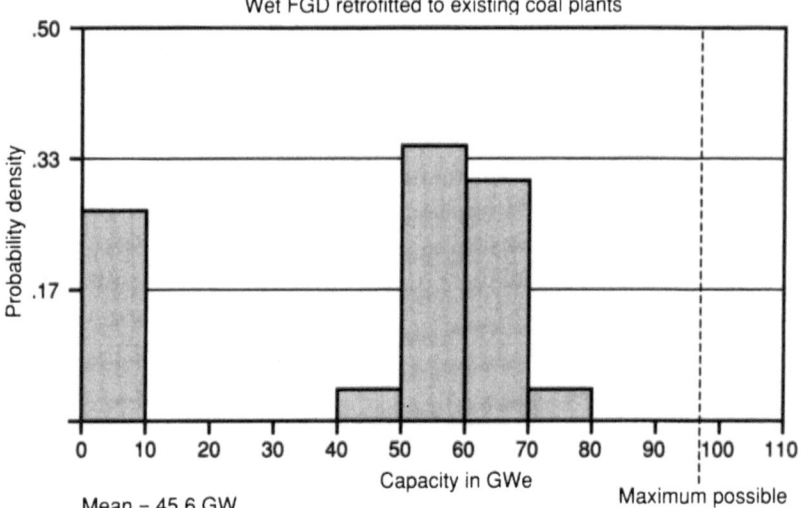

Panel E

Wet FGD retrofitted to existing coal plants

Mean = 45.6 GW

Figure 5-1. *(continued)*

The above observations suggest that AFBC investment would emerge as a cost-minimizing strategy in the model only under technology assumptions that are significantly different than the assumptions used in this study. That is certainly possible. The expected performance or cost of AFBC relative to IGCC could be more favorable than reported in the EPRI *Technical Assistance Guide*. Similarly, AFBC could be found to have a lower probability of high cost than IGCC.

To further investigate the various factors determining the attractiveness of AFBC investment, we undertake a model run with a larger AFBC unit size—500 MW, as compared with the 200 MW size assumed in other runs. The differences in assumptions for the two AFBC units are summarized in table 5-1. The larger unit size has a lower unit capital cost, reflecting the presence of some scale economies in AFBC construction, and a lower O&M cost. However, it also has lower availability. All other inputs are set at their modal values.

The results of this model run are summarized in table 5-2. Some AFBC investment does occur in this case, indicating that lower unit capital costs offset lower unit availability. However, the size of the investment is still quite small. This result is particularly interesting, since the assumed capital cost of the 500 MW AFBC unit is competitive with the capital cost of pulverized coal/flue gas desulfurization (PC/FGD) technology. Thus this experiment suggests that for AFBC to be a preferred outcome in the model it must have fundamentally

Table 5-1. Properties of 200 MW and 500 MW AFBC Units

Property	Unit size of AFBC plant	
	200 MW	500 MW
Capital cost (1984 $/kW)[a]	1,508	1,096
O&M cost (1984 mills/kWh)	7.8	6.5
Heat rate (Btu/kWh)[b]	10,045	10,000
Operating availability (%)	81.5	74.4

[a] Represents "overnight construction" cost.
[b] Represents average annual heat rate.
Source: Electric Power Research Institute (1986b).

Table 5-2. Impact of Economies of Scale in AFBC Units on New Capacity Investment in Regions E and M in 2010

Technology	Unit size of AFBC plant	
	200 MW	500 MW
	(GWe)	(GWe)
AFBC	0.0	1.5
GTCC	14.5	14.5
IGCC	35.8	34.8
PC/FGD	2.4	2.2
Total	52.7	53.0

superior technology and cost characteristics than are presented in the EPRI *TAG* report.

In a second experiment, we reverse the risk classifications that are assigned to IGCC and AFBC in chapter 4, treating the former as a more risky pilot technology and the latter as a demonstration technology, while holding other inputs at median values. This reversal changes the probability distributions for IGCC and AFBC construction costs, following the procedures sketched in chapter 4 and table 4-3. The alternative distributions are summarized in table 5-3.

The impact of these changes on capacity investment decisions is summarized in table 5-4. These results show that the preference in the model for IGCC over AFBC is not due to our assumption about risk classifications. The change in risk classifications does lower the expected level of IGCC investment, but under the maintained assumptions it is PC/FGD rather than AFBC that becomes more attractive for new investment. Again we are led to the conclusion that the cost and performance characteristics of AFBC must be fundamentally superior to those listed in table 4-1 for this technology to be considered a preferred option in the model.

Returning now to figure 5-1, we note other interesting characteristics of the technology choice distributions. The distribution of GTCC

Table 5-3. Alternative Construction Cost Distributions for AFBC and IGCC Units

	Costs ($/kW)[a]		
Case[b]	Minimum	Mode	Maximum
AFBC more risky			
AFBC cost	1,206	1,508	2,412
IGCC cost	1,199	1,411	1,987
IGCC more risky			
AFBC cost	1,311	1,508	2,111
IGCC cost	1,128	1,411	2,258

[a] All figures are 1985 dollars.
[b] See text, chapter 5, for definitions of these cases.

Table 5-4. Levels of Investment Under Alternative Assumptions About AFBC and IGCC Construction Cost

	Investment levels (GWe)	
Technology	AFBC more risky	IGCC more risky
AFBC	0.0	0.0
GTCC	14.5	18.0
IGCC	35.8	23.4
PC/FGD	2.4	12.8
Total	52.7	54.4

investment (figure 5-1, panel A) is almost uniform over the 0–30 GW range (the largest observed value is just over 30 GW and carries a low probability). The distribution of IGCC investment (panel B) is spread over a wider range (0–70 GW), indicating greater uncertainty about baseload capacity requirements, and it is more asymmetric at the lower end. This asymmetry reflects the possibility that new PC/FGD units could be a lower-cost alternative. However, investment in new PC/FGD units is modest except in a few low-probability cases (panel C). In the model, PC/FGD units have longer lead times and thus higher actual construction costs than IGCC units. In addition, the FGD units add significantly to basic operating costs for new PC plants, as noted in chapters 3 and 4.

The shape of the distribution for total capacity investment (panel D) is also roughly uniform. This shape is like that of the input distribution for demand growth, and contrasts with the shapes of other input distributions (for fuel prices, plant costs and performance), which are single-peaked. This suggests that demand may be the most important influence on investment in the model results. We investigate that possibility further in the sensitivity analysis.

Panel E of figure 5-1 presents results for investment in FGD retrofits of existing coal plants. Roughly 99 GW of plant capacity in regions E and M is potentially a candidate for retrofitting. Panel E shows that in roughly one-quarter of the model runs, no more than 10 GW of retrofitting is actually undertaken. These runs correspond to imposition of only the lightest constraint on sulfur emissions (case 1, table 4-7). In the other runs FGD retrofitting is significant, ranging from 40 GW to 80 GW of capacity. The fact that in this part of the histogram there are not just three equiprobable bars corresponding to the other three emission scenarios (cases 2–4 in table 4-7) indicates that factors in the model other than environmental constraints also have a significant impact on FGD retrofit investment. For example, FGD investment is lower (40–50 GW) in relatively unlikely cases involving very low demand growth and IGCC costs; similarly, it is higher (70–80 GW) with high demand growth and IGCC costs.

One other experiment of interest relates to the choice of discount rate used for calculating present values in the model. As noted in chapter 4, our main set of model runs includes different settings of the real discount rate between 6 and 12 percent. As large as the upper end of this range is, it could be argued that a still higher rate is appropriate in practice given the higher financial risks faced by electricity generating companies relative to historical norms.[3] To address the effects of a high discount rate, we undertake a special model run with an 18 percent (real) rate, holding other inputs at modal values.

The resulting investment pattern is compared to results with a 9 percent rate in table 5-5. We would expect a higher discount rate to shift investment away from technologies having higher capital costs. Table 5-5 reveals precisely this pattern, with a shift in investment to GTCC. IGCC investment still is substantial, however, while PC/FGD investment disappears and AFBC investment remains at zero. These results reflect the longer construction times for the latter technologies, and thus the greater sensitivity of their costs to the discount rate.[4]

---

[3] For utilities these higher financial risks may arise from higher regulatory and economic risks. For new nonutility generating companies, higher financial risks reflect higher economic risks and lower regulatory protection.

[4] Though the model treats GTCC and IGCC investments as separate events, the results in table 5-5 can be interpreted as indicating that a high discount rate induces a slower conversion of GTCC units to IGCC units. This view is bolstered by the observation that a higher discount rate works to offset rises in gas prices relative to coal prices, thus strengthening the position of GTCC relative to IGCC in present-value terms.

Table 5-5. Impact of High Discount Rates on New Capacity Investment in Regions E and M

	Investment levels (GWe)	
Technology	9% real discount rate	18% real discount rate
AFBC	0.0	0.0
GTCC	14.5	29.0
IGCC	35.8	23.1
PC/FGD	2.4	0.0
Total	52.7	52.1

Impacts of Different Power Trading Regimes

Expanded interstate power trading should reduce the total economic cost of power supply for regions E and M by enlarging the range of options available for meeting demand.[5] Table 5-6 shows the ''average marginal costs'' of in-state power generation for the nine states in the year 2010 (the terminal date of the model calculations), with and without new trade. The entries in the table are weighted averages of marginal generation costs across each block of the load duration curve, with weights equal to the relative durations of the blocks (see figure 2-3).[6] The figures in table 5-6 reflect mean values of the outputs obtained from the 60 runs of the model.

Table 5-6 shows that, as expected, expanded trading options in the model work toward reducing differences in marginal generation costs among the states in region E. States that are net exporters—here, E2 and E5—have higher marginal costs as system capacity factors increase to meet export demand as well as in-state demand. Correspondingly, the net importing states—here, E1, E3, and E4—show a decline in marginal supply cost as in-state generation is displaced by lower-cost imports.

Marginal costs among the region M states are already drawn together in either trading regime by the central dispatching of the pool's capacity. However, expanded trading options in the model do produce a slight drop in marginal generation costs for each region M state, indicating that region M acts in the model as a net importer (beyond existing contractual volumes) from region E.

In terms of actual trading volumes, detailed model results (not presented here) indicate that E5 is the largest exporter, while E4, M2,

---

[5] It should be recalled that all the trading options under consideration refer to long-term supplies, not short-term coordination transactions.

[6] Thus the averages include incremental capital costs for demand blocks in which capacity additions are undertaken, but for blocks in which no capacity additions are undertaken they include only fuel, operating, and maintenance costs.

Table 5-6. "Average Marginal Cost" of Electricity Generation in 2010 Under Different Trade Assumptions (mills/kWh)

State	No new trade	New trade possible
E1	33.16	29.14
E2	25.83	30.24
E3	37.72	33.56
E4	36.05	35.64
E5	21.87	28.50
M1	39.95	39.91
M2	39.41	39.28
M3	38.71	38.68
M4	34.73	34.10

Note: The cost figures are in 1985 dollars and are the average of marginal costs over all blocks of the load duration curves. They do not include the capital depreciation of existing plants, or the costs of spinning reserves. The scenarios are defined in the text.

and M4 are the largest importers. Table 5-7 illustrates the size of the supplier premium paid by each of these importing states when purchasing power from E5. Column (1) shows the average marginal cost of in-state generation with no new trading; column (2) shows the average marginal cost of power generation and transmission from E5 to the importing state under full coordination (with no trading premium built into the cost). The potential cost savings from trade is the difference between the figures, shown in column (3). This difference is significant—roughly 20 percent of the in-state generation cost with no new trade. Half of the difference, shown in column (4), is the amount added to marginal generation plus transmission costs to determine the delivered cost of imports when new trade is allowed.[7]

Different trading options are also expected to affect the size and composition of new capacity investment. In figure 5-2 (pp. 63–65), panels A through E contain probability histograms (analogous to panels A–E in figure 5-1) in which we assume no new trade possibilities among the region E states and the region M pool. In general, the lack of new trading options raises new capacity investment requirements for each technology. The new investment distributions are also somewhat less uniform than in figure 5-1, suggesting an increased importance of influences other than demand. Generally, however, the new capacity investment distributions do not radically differ from the corresponding results with new trading options. The

---

[7] This cost will differ slightly from the sum of costs in columns (2) and (4) in table 5-7 because under the regime where individual states can pursue new trade options, E5 will dispatch plants differently than it would under full coordination; hence its marginal generation plus transmission costs will differ from those in column (2).

Table 5-7. Projected Premiums for Power Imports in 2010 (mills/kWh)

Importing state	Generation Cost[a]		Difference	Supply premium
	In-state[b] (1)	Imported[c] (2)	(3)=(1)−(2)	(4)=0.5×(3)
E4	36.05	28.98	7.07	3.54
M2	39.41	32.03	7.38	3.69
M4	34.73	26.79	7.94	3.97

[a] Figures are "average marginal costs" as defined in table 5-1.
[b] Assumes no new power trading.
[c] Sum of generation and transmission costs from state E5, assuming full coordination. Transmission losses, which are dependent on the distance of transmission, are covered by the importing state.

same is true for the distribution of FGD retrofit investment (figure 5-2, panel E), except that we observe a leftward shift of 10 GW in the cases involving binding environmental constraints. This points to increased reliance on generation from older plants, and thus a greater need for retrofits, with expanded interstate trade.

Table 5-8 (p. 65) summarizes the differences in total investment in regions E and M across the two trading regimes, expressed in terms of averages of results over the 60 model runs. New capacity investment in each technology is reduced by the provision of enhanced trading opportunities. Total new investment drops by about 13 percent; investment in GTCC, mainly for meeting peak demands, drops almost 20 percent. These results complement those given in table 5-7 in showing the potential cost savings yielded by expanded trade. However, these gains might not be realized in practice since we have ignored possible impediments to expanding transmission capacity.

Sensitivity Analysis

We summarize here some findings from our sensitivity analysis under the assumption that new interstate power trading can occur. Findings for region E and region M are presented in tables 5-9, 5-10, and 5-11 (pp. 66, 67). For each output considered in these tables we identify the input assumptions that cause the most variation across model runs in the model outcome, as revealed by rank correlations between inputs and outputs. Only inputs with a significance level of 5 percent or greater are listed in the tables.[8]

---

[8] In a few cases we have omitted a factor that nominally met this test, where the result stems from colinearity of inputs in the sample and clearly is nonsensical. This illustrates the need for care in interpreting the results of the sensitivity analysis, given the dangers of spurious correlation.

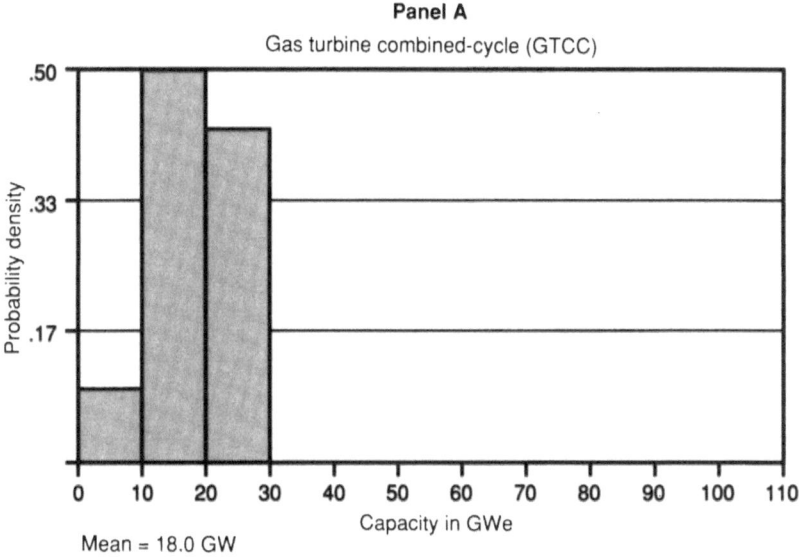

Panel A

Gas turbine combined-cycle (GTCC)

Mean = 18.0 GW

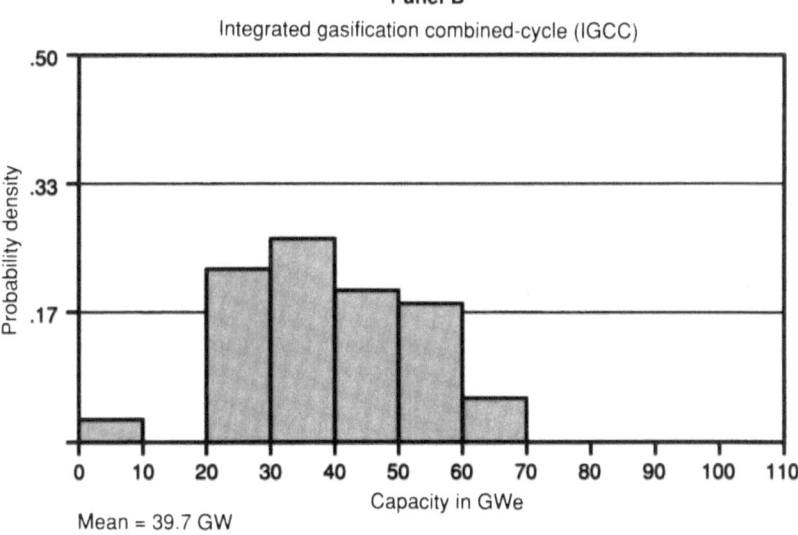

Panel B

Integrated gasification combined-cycle (IGCC)

Mean = 39.7 GW

Figure 5-2. Investment histograms when no new interstate trade is allowed

(continued)

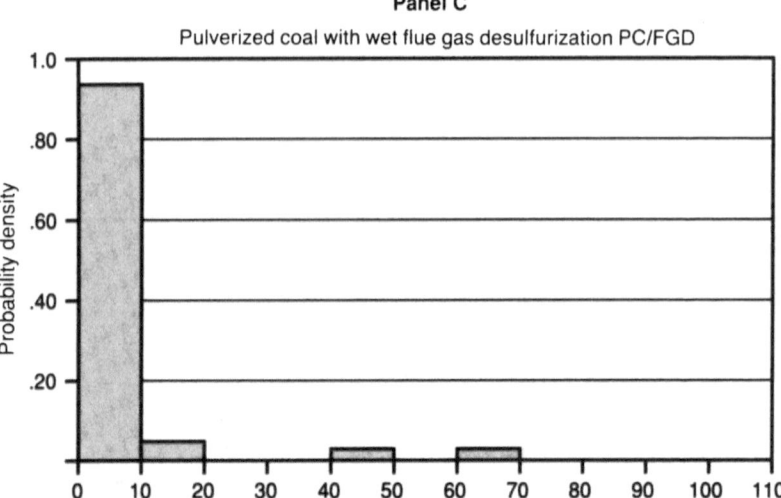

Panel C

Pulverized coal with wet flue gas desulfurization PC/FGD

Mean = 2.8 GW

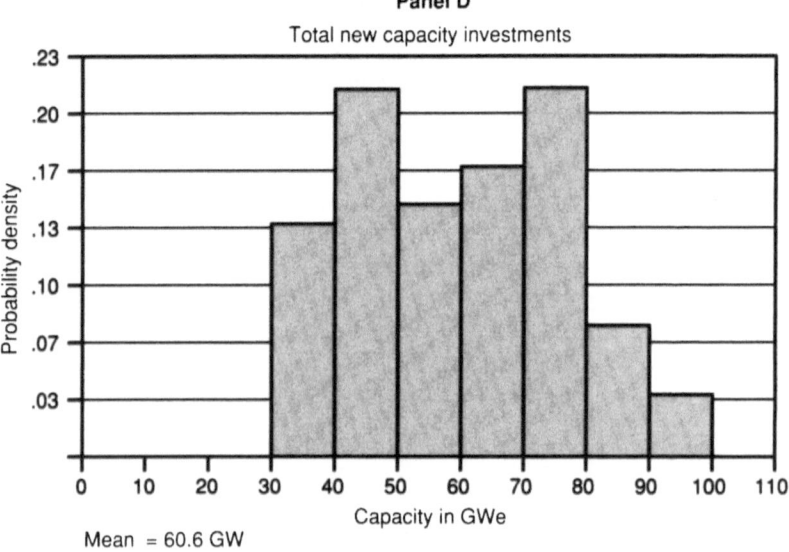

Panel D

Total new capacity investments

Mean = 60.6 GW

Figure 5-2. *(continued)*

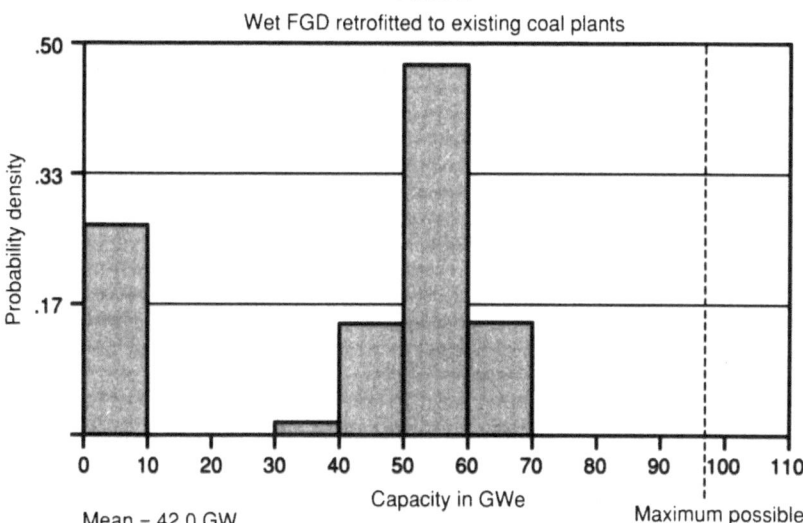

Panel E

Wet FGD retrofitted to existing coal plants

Mean = 42.0 GW

Maximum possible

Figure 5-2. *(continued)*

In tables 5-9, 5-10, and 5-11, the row numbers (1 through 3) indicate the ranking of the model inputs in terms of their explanation of observed variation in the corresponding model outputs. The plus or minus sign following each entry indicates the direction of the relationship between that input and the output in question. In tables 5-9 and 5-10 the model outputs are indicated by the column headings; they refer to different technology choices for new investment and to FGD retrofits for existing coal plants. Thus, for example, table 5-9 shows that the top three influences on GTCC investment in region E are (in order of importance) the overall growth in electricity demand, success in load management (LM), and the cost of IGCC investment. Demand growth and higher IGCC cost raise GTCC investment, while

Table 5-8. Expected Total New Capacity Investment Required in Regions E and M by 2010 Under Different Trade Assumptions

Technology	No new trade (GWe)	New trade possible (GWe)
AFBC	0.0	0.0
GTCC	18.0	14.5
IGCC	39.9	35.8
PC/FGD	2.8	2.4
Total	60.6	52.7

Table 5-9. Rank of Significant Factors Influencing Investment in New Capacity in 2010

Rank of input in region E	GTCC	IGCC	PC/FGD
1	Demand (+)	Demand (+)	IGCC cost (+)
2	LM (−)	LM (−)	PC HR (−)[a]
3	IGCC cost (+)	. . .	. . .
Rank in region M			
1	Demand (+)	Demand (+)	IGCC cost (+)
2	LM (−)	LM (−)	PC cost (−)[b]
3	IGCC cost (+)	. . .	Demand (+)

Note: Factors listed are significant at the 5 percent level in nonparametric tests. Demand = growth in electricity demand; IGCC cost = unit construction cost for integrated coal gasification combined-cycle plants; LM = success in load management programs.

[a] Heat rate of pulverized coal power plants.

[b] Unit construction cost for pulverized coal plants.

Table 5-10. Rank of Significant Factors Influencing Investment in FGD Retrofits in 2010

Rank of input in region E	SIP plants[a]	NSPS plants[b]
1	Demand (+)	Demand (+)
2	IGCC cost (+)	. . .
3	. . .	. . .
Rank in region M		
1	LM (+)	LM (+)
2	IGCC cost (+)	GTCC HR (−)[c]
3	O&G price (−)[d]	Environmental regulations (+)[e]

Note: Factors listed are significant at the 5 percent level in nonparametric tests. Demand = growth in electricity demand; IGCC cost = unit construction cost for integrated coal gasification combined-cycle plants; LM = success in load management programs.

[a] Investments in retrofit FGD equipment for older (State Implementation Plan) coal-fired power plants.

[b] Investments in retrofit FGD equipment for newer (New Source Performance Standard) coal-fired power plants.

[c] Heat rate of gas turbine combined-cycle plants.

[d] Price of oil and gas used for generation of electricity.

[e] Severity of environmental regulations for control of SO_2.

success in load management lowers GTCC investment. The input rankings in table 5-11 refer to the average marginal cost of electricity generation in each of the nine states included in the study. For example, the top three influences on the cost of electricity generation in state E1 are demand growth, IGCC cost, and the discount rate; increases in these factors all raise generation cost.

Table 5-11. Rank of Significant Factors Influencing the Average Marginal Cost of Electricity Generation in 2010

Rank of input	State				
	E1	E2	E3	E4	E5
1	Demand (+)	Demand (+)	Demand (+)	Demand (+)	Demand (+)
2	IGCC cost (+)	Discount rate (+)	IGCC cost (+)	IGCC cost (+)	⋯
3	Discount rate (+)	⋯	O&G price (+)	O&G price (+)	⋯

Rank of input	State			
	M1	M2	M3	M4
1	Discount rate (+)	Discount rate (+)	Discount rate (+)	Discount rate (+)
2	IGCC cost (+)	IGCC cost (+)	IGCC cost (+)	IGCC cost (+)
3	O&G price (+)	O&G price (+)	O&G price (+)	Demand (+)

Note: Factors listed are significant at the 5 percent level in nonparametric tests. Demand = growth in electricity demand; IGCC cost = unit construction cost for integrated coal gasification combined-cycle plants; O&G price = price of oil and gas used for generation of electricity.

Before examining the specific results in these tables, it is useful to further explain in general terms the information that the sensitivity analysis conveys. By construction, the rank correlations reveal the input parameters whose variations in the stochastic sample are most strongly related to the observed variations of outputs generated by the model. These parameters are likely also to strongly affect the outputs in an absolute sense, but the converse is false: there may be inputs that are very important influences on the outputs, in the model and in practice, but that are not so identified by the sensitivity analysis. This can happen particularly when the probability distributions for some inputs are concentrated on values where there is little effect on a model output, while excluding other parameter values where the relationship with the output is stronger. Examples of this situation are provided below. We should also note that the sensitivity analysis illustrates how outputs for a given year—2010, the end of the assumed time horizon—vary with input assumptions across model runs. These are different relationships than the responses of model outputs to fluctuations in inputs over time.

Turning to the actual results, the ranking of different factors influencing new capacity investment is shown in table 5-9. Many of these results are fairly intuitive. Since the bulk of new investment in our runs is in GTCC or IGCC units, both of these decisions are significantly affected by the level of demand growth and the success of load management programs (the minus sign after the LM entries indicates that less success in these programs implies higher investment requirements). GTCC investment also is a partial substitute for IGCC in cases where IGCC cost is higher.

In contrast, PC/FGD investment occurs more sporadically in the model runs (see figure 5-1, panel C); variations in PC/FGD investment are influenced primarily by the cost of the competitive alternative in the model—IGCC—and by its own performance characteristics, rather than being heavily influenced by demand. We can combine these observations to illustrate conditions under which GTCC and IGCC may not be the dominant new investments in the model. If demand growth is sluggish, load management programs are successful in flattening load duration curves, and IGCC construction costs are high, then total new investment will be lower but the share of PC/FGD units in new investment will be higher.[9]

One other result of interest in table 5-9 is that the factors indicated to be most strongly correlated with new investment are very similar for both region E and region M, even though these regions differ

---

[9] This will also apply to AFBC investment if other conditions are favorable to the presence of this technology in the new investment mix.

significantly in their existing plant-base and demand characteristics (see tables 2-1 and 2-2). This suggests that our conclusions about technology choice are not a fluke, even though they clearly do depend on our basic costs and performance assumptions.

Two sets of influences are conspicuous by their virtual or total absence from table 5-9: technology cost and performance characteristics, and fuel prices. The near absence of the former is partly explained by our earlier argument concerning the lack of AFBC investment in the results: our assumptions make IGCC the cost-minimizing baseload technology of choice in most situations, so there is relatively little variation in the model results between the cost or performance of coal-fueled technologies and the technology choices indicated by the model. Such variation would show up with wider ranges of input assumptions for cost and performance. This illustrates how the results of the sensitivity analysis are shaped by the nature of our input assumptions.[10] Similarly, our fuel price assumptions have no significant impact on the choice among coal technologies, and our assumptions about gas prices relative to coal do not lead to significant variations across model runs in competition between GTCC and IGCC for new baseload capacity (this effect is dominated by the impacts of variations in IGCC capital costs). The price of gas would appear in table 5-9 only if we considered very low gas price growth rates that would keep the two fuels more cost-competitive, given the other properties of the technologies.[11]

Table 5-10 summarizes the main influences in the model on FGD retrofit investments for two vintages of existing coal plants. Here again we find that demand factors are the primary source of variation across the model runs, but the precise influences differ between region E and region M. Variations in the growth in demand for region E power (including imports by region M) have a significant effect on unit capacity factors in region E. When demand growth and capacity factors are moderate, environmental constraints can be met most cheaply by switching to higher-priced low-sulfur coal; when demand growth and capacity factors are high, the more capital-intensive option of FGD investment becomes the preferred option. This chain of reasoning explains why FGD investment in region E is significantly influenced by total demand growth in the model.

---

[10] Nevertheless, our analysis covers a wider range of cases than studies that treat only a couple of scenarios.

[11] The observed importance of fuel prices would also be greater if the model included staged investment in IGCC (with a GTCC first stage), and if a larger number of fuel types were considered in new capacity investments. In addition, if we considered time paths of investment and allowed for conversion of GTCC to IGCC, we would find that the timing of this conversion is sensitive to gas prices.

In region M, by contrast, current reserve margins are relatively slim (see table 2-2), and just about any assumed demand growth rate will result in capacity factors sufficiently large to favor FGD investment. This is another example of how the apparent absence of a significant influence in the sensitivity analysis results can be explained by the nature of our input assumptions: we would only see total demand growth significantly influencing FGD investment in region M if we allowed for (perhaps implausibly) low demand growth. On the other hand, load management is a significant influence in region M because flattening-out the load duration curves in region M states would reduce the need for incremental combined-cycle capacity and raise the capacity factor of existing coal plants, thus stimulating FGD investment.

Significant influences on the average marginal cost of generation for all nine states considered in the study are summarized in table 5-11. Once again we find that changes in the demand growth assumption are a major source of variation in the results for the states in region E. More rapid demand growth in these states raises the capacity factors of all units, including higher-cost units, as well as necessitating more new investment, thus raising the cost of generation.

In region M significant new capacity investment occurs over all demand scenarios in the model. The biggest influence on the cost of this investment, and thus on the average marginal cost of generation, is found to be the discount rate. In both regions, IGCC capital cost is a significant influence on marginal generation cost, given the model's frequent choice of this technology for new investment. Finally, oil and gas price growth is a significant influence in states having a significant number of existing oil-fired steam units, and where significant quantities of gas will be used for GTCC plants in the future.

Summary: Strengths and Weaknesses of the Model Results

Several important conclusions are indicated by our model results.[12] The first is the general favoring of advanced modular plant designs over conventional coal technology for new plant investment. This result does not depend solely on environmental constraints, though the advanced designs are capable of achieving relatively low emis-

---

[12] We stress again that the outputs of the model could not be viewed as specific forecasts or recommendations without increasing the range of technology options considered and expanding the degree of detail in the databases on existing plants, among other things.

sions; advanced designs are preferred even when there are less stringent sulfur oxide limits that can be met with FGD investments for new plants and retrofits. While this finding applies only to new investment, it suggests that over time the capital stock of the electricity generation industry will look very different than it has in the past.

The favoring of gas- or coal-fueled combined-cycle units over fluidized bed units in new investment is also striking, and illustrates the value of our stochastic sensitivity analysis procedures. In 60 model runs no AFBC investment is observed, even though the model is biased against IGCC units because it ignores conversion of GTCC units. We also find that our conclusions about the composition of new investment do not appear to be greatly sensitive to assumptions about new interstate power trading. Expanded trade does seem to offer opportunities for significant overall cost reductions and cuts in total investment required to meet future electricity demand, but these conclusions depend on the capability to expand transmission.

In general, the model results accord with intuition concerning how investment size and technology choice vary with different influences (electricity demand, supply costs, environmental constraints, and the nature of existing plants). Where this is not the case, an explanation usually can be found in the nature of the input assumptions. The results on how expanded interstate power trading could reduce aggregate supply costs and investment requirements are also intuitively reasonable, though the model is silent on how these potential gains might be thwarted by transmission constraints.

The finding that in the model advanced technologies are preferred over conventional coal plants with FGD may be somewhat controversial. Different assumptions about technology cost and performance could lead to different results. Nevertheless, the finding can be defended given the nature of the technologies in question, as pointed out in chapter 3. The modular nature of advanced-technology units allows economies in production and partial assembly at the plant site, which can offset the loss of scale economies from a smaller unit size. The smaller unit scale also facilitates a closer matching of supply increments with demand growth. Moreover, the advanced technologies considered in this study appear to have greater thermodynamic efficiency, smaller pollution streams, and lower total costs (including the cost of pollution abatement) than incremental investment in conventional plants.

Our finding that the model chooses no investment in AFBC is probably the most controversial result, since a different set of technology assumptions than ours (which are derived partly from EPRI's *Technical Assistance Guide*) could lead to different conclusions. A more complete treatment of the subject would involve consideration of a wider range of assumed technology characteristics, but such an effort is beyond the scope of the present study.

We should point out again, however, that the option of IGCC investment chosen by the model does offer several advantages not reflected in the PADRE framework, so our results favoring IGCC are to some extent conservative. Staged construction, starting with a GTCC unit, lowers the present value of capital cost and provides strategic flexibility in responding to relative fuel prices. The staged timing of capacity investment also provides greater flexibility in responding to different short-term and long-term contractual opportunities in increasingly competitive bulk-power markets.[13] Thus the basic thrust of our results on IGCC and AFBC investment cannot simply be rejected out of hand. Clearly, additional investigation of the relative merits of these technologies is needed.

We also note once more that there are inherent limits to using our engineering-economic approach for gauging how investment choices in the generation sector may actually be made. One drawback, it will be recalled, is the limited ability of our approach to address issues related to industrial organization and economic regulation, including the effects of burgeoning competition. The model also cannot incorporate influences related to long-term contracting and other institutional factors, such as siting of new plants or transmission capacity. To illustrate, many utilities obtained land, siting permits, and licenses for additional plants in the 1970s but then postponed or canceled planned expansions. Questions of site access could influence which companies are most likely to expand capacity, if not the overall level of capacity expansion.[14]

Another important drawback is the limited ability of the model to address how uncertainty might affect decisions by utilities. While our sensitivity analysis embodies subjective probability distributions for key inputs, each execution of PADRE is deterministic—there is no uncertainty built into the cost-minimizing calculations themselves. Therefore the model does not address how ex-ante uncertainty affects the path of investment decisions, particularly if utility planners are risk-averse. The sensitivity analysis deals only with random ex-post variation of input assumptions.

While the model does not deal directly with risk aversion, the outputs can be interpreted in ways that shed some light on reactions to

---

[13] Specifically, capital cost risks in short-term contracts can be mitigated by delaying the conversion from GTCC to IGCC; fuel-cost risks in long-term contracts can be mitigated by moving sooner to coal.

[14] Utilities could lease or sell sites to other companies, or form joint ventures in lieu of expanding capacity themselves.

technology risks. Risk-averse investors will not simply minimize expected costs in choosing new capacity increments, as postulated in the model. Instead, they can be expected to trade off reductions in expected costs against reductions in risk.[15] One simple rule for doing so is to associate with each uncertainty in the cost stream a risk premium commensurate with the uncertainty and the cost of risk-bearing. For example, if (hypothetical) technology A had no uncertainty in its construction costs while the costs for technology B were uncertain, a premium could be added to the mathematical expectation of B's costs for purposes of comparison with A's costs. The decision maker could then be viewed as minimizing expected costs after incorporating the risk premiums (see Lind, 1982).

Our technology assumptions reflect the relative risks associated with the performance and costs of different technologies in two ways. Conventional generation technologies are assumed to have the least uncertainty in ex-ante estimates of their cost and performance parameters, while advanced technologies are assumed to have a higher degree of uncertainty in these parameters. In addition, the uncertainty bounds for advanced technologies are asymmetric, with modal values equal to EPRI figures and long upper tails in the distributions. This raises mean unit construction costs for these technologies and lowers their mean thermodynamic efficiencies. This in turn is similar to adding a risk premium to the capital costs of the advanced generation technologies.[16]

Experience clearly matters in deploying new technologies, however.[17] The pattern of technology ownership could vary across the electricity generation industry. For example, since utilities generally have less experience with managing a chemical process like coal gasification, whereas other firms (such as petrochemical companies) do have this experience, there could be a divergence in technology ownership among utilities and independent generating companies within the overall pattern of investment. Alternatively, utilities and other firms could form joint ventures for IGCC plants, in which the utility would maintain and operate the combined-cycle plant and the other firm would be responsible for the gasification of coal. In this example the model's limitations in addressing institutional issues are evident.

---

[15] See Rose and Joskow (1988) for a discussion of this behavior in the context of electricity generation.

[16] Such premiums can be quite substantial. The premium for IGCC plants is close to $100/kW, while that for AFBC plants is about $150/kW.

[17] For empirical evidence, see Joskow and Rose (1985), Joskow and Schmalensee (1985), and Rose and Joskow (1988).

Despite these caveats, the model results suggest that our methodology is useful for its designed purpose—calculating economic cost-minimizing technology choices. As we note at the outset, the analytical framework brings together many important elements of the issue: investment planning over time, interactions among multiple supply regions, and selection from among a number of options for new plant or retrofit investments. The stochastic sensitivity analysis helps to promote objectivity in the assessment of technology options by allowing a wide range of input assumptions to enter the analysis, by indicating how model outcomes are shaped by input assumptions, and by highlighting technology choices that appear to be desirable under a number of different market conditions.

6

ENVIRONMENTAL POLICY DESIGN AND TECHNOLOGY CHOICE

In this chapter we illustrate the application of the model to policy issues, with specific reference to issues raised by the 1990 revisions of the U.S. Clean Air Act.[1] The sensitivity analysis presented in chapter 5 indicated that for the ranges of input parameters considered, the *severity* of limits on SO_2 emissions was not an important influence on technology decisions. The revisions of the Clean Air Act make possible new ways in which utilities can meet requirements for SO_2 reduction. It is therefore interesting to examine briefly whether the *design* of SO_2 regulations might significantly affect technology choice.

We address environmental policy design for the five states that constitute region E. These states face more serious air pollution problems from power plants and have less interstate coordination than the four states that make up the region M power pool. It is important to keep in mind the limited scope of our analysis. The Clean Air Act revisions are national in scope and allow many more possibilities for interstate tradeoffs in emissions control than our analysis encompasses. Consequently, our results must be seen as illustrative only.

Before the 1990 revisions, a command-and-control approach was used for regulating SO_2 emissions under which each plant was required not to exceed a certain level of emissions (expressed in terms of pounds of SO_2 per million Btu of energy input). The revisions

---

[1] Because our purpose is illustrative, we keep the discussion in this chapter brief. Readers interested in additional details should consult Dowlatabadi and Harrington (1989, 1990), which we draw upon extensively.

establish a multistate tradeable permit regime, in which each plant would be given a certain base volume of emission rights that it could use, sell to other utilities, or supplement with purchases of additional allowances. Plants having below-average abatement costs would undertake more emission reductions, overcontrolling to generate a valuable asset in the form of additional allowances that could be sold to other plants having above-average abatement costs. These economic incentives are expected to move the industry closer to a least-cost total allocation of abatement effort, in which marginal abatement costs are equal across emitters, while avoiding some of the oversight burden associated with command-and-control.[2]

We consider here four different policy regimes for SO_2 control: (1) a command-and-control approach with plant-specific emission standards; (2) a marketable permit policy allowing interstate trades among power plants; (3) a marketable permit policy allowing only intrastate trades among plants (so that there is a separate permit market within each state); and (4) a tax on SO_2 emissions. Option (2) most closely resembles the 1990 Clean Air Act revisions, while option (3) is essentially the approach used in this analysis so far, though we have not explicitly postulated intrastate emission trading. Option (4) is another method that relies upon economic incentives and involves only limited centralized oversight of individual utility decisions. Like tradeable permits, the tax has the potential for engendering cost-effective emissions control. Each emitter would control only to the point where the marginal cost of abatement equals the tax, so marginal abatement costs are equalized across emitters.

The most important differences between tradeable permits and the emissions tax is that with the former, total emissions are given by construction (ignoring cheating), while with the latter the regulators probably would need to experiment with different tax rates to induce the desired total emissions level. This difference becomes important when one recognizes regulators' inherent uncertainty about future market conditions when the emissions tax is set. If the evolution of demand, of cost and performance of different generation technologies, as well as other factors, were known with certainty, taxes and permits could be used to the same end.

With uncertainty about demand and technology, however, regulators must commit to a tax rate before the ''state of the world'' is known. Suppose they choose a tax rate that yields a desired level of emissions given some particular expectation of future economic and

[2] For a discussion of emissions trading and the Clean Air Act revisions, see Bohi, Burtraw, Krupnick, and Stalon (1990). Tietenberg (1985) provides a general treatment of marketable permits for pollution control.

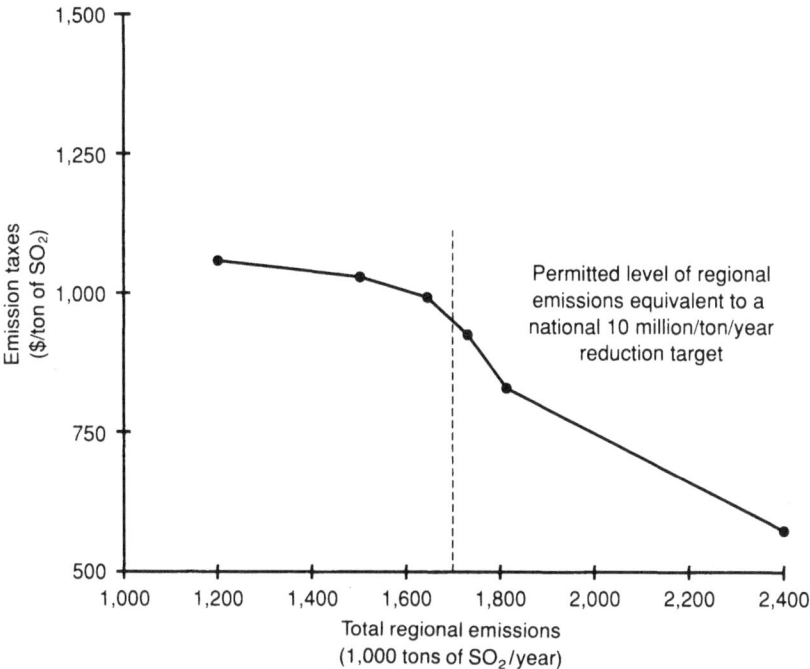

Figure 6-1. Calculation of an emissions tax for Region E. All input parameters are set equal to their expected values.

technological conditions. With this fixed tax, the actual level of emissions will depend on the state of the world but the marginal cost of abatement is known (it is equal to the tax). In contrast, with permits the emissions level is determined with certainty (assuming full compliance) but the level of abatement cost is uncertain given uncertainty about supply and demand conditions.

Figure 6-1 illustrates the calculation of an emissions tax for the five states in region E when regulators' expectations about future conditions are given by the mean values of the input parameter distributions discussed in chapter 4.[3] As one would expect, a higher tax implies a lower expected volume of emissions. An emission constraint of 2 pounds of SO_2 per million Btu of plant input energy—which corresponds to the national emissions reduction of 10 million tons below 1980 levels called for in the Clean Air Act revisions—

[3] It should be recalled that in chapter 5 we frequently made use of model runs with parameters set at their modal rather than their mean values. Our interest here, as developed below, is in comparing outcomes under the mean tax rate with the mean of outcomes under a permit strategy.

results in a region E emissions reduction target of about 1.7 million tons per year, a 4.5 million ton reduction from 1980 levels. To induce this response with the input parameters at their mean values, a tax of roughly $850 per ton of SO_2 emissions would be necessary.

Results from the Model

To address the issues raised above, we again exercise the model for 60 different sets of input parameters, using the stochastic sampling procedure discussed in previous chapters. For each of the runs we consider the four different environmental policy designs enumerated above—fixed plant standards, interstate permit markets, intrastate permit markets, and an emissions tax. The standard and permit cases are carried out assuming a 2 pound per million Btu SO_2 limit on power plants (this is an absolute limit in the standard case, but a benchmark that can be altered for individual plants with tradeable permits). The tax case uses the $850/ton figure developed in figure 6-1, which is the tax that meets the 2 pound per million Btu aggregate limit given expected values of the input parameters. One interesting finding is that with this tax, actual emissions were within ±15 percent of the target over all 60 runs of the model. At least for the area we are considering and the assumptions we have made, uncertainty about emissions does not seem to be an overwhelming drawback of the emissions tax.

Table 6-1 summarizes some of the variability of the technology choices across the 60 model runs for each of the four policy designs. The top half of the table considers flue gas desulfurization (FGD) retrofits to existing plants, while the bottom half considers investments in new clean-coal technologies.[4] The FGD results show that a rigid performance standard increases FGD retrofits by roughly a third over the more cost-effective alternatives, essentially because the inflexibility of the standard does not allow abatement to be allocated cost-effectively. The other main point conveyed by this part of the table is the small differences in retrofit activity across the incentive-based policy designs. This is due partly to the limited differences between emissions under a tax approach and a permit approach. There also appears to be some homogeneity among the states in region E that makes the benefits of interstate permit trading over intrastate trading modest (the benefits of FGD retrofits with the latter policy are only slightly larger than with the former).

---

[4] As in chapter 5, this investment is primarily if not exclusively in integrated gasification combined-cycle (IGCC) plants in lieu of fluidized bed plants.

Table 6-1. Investment in Pollution Control and Clean-Coal Technologies Under Four Policy Designs

Policy design	Investment in FGD retrofits to existing coal-fired power plants (GW)		
	Minimum	Mean	Maximum
Performance standard[a]	47.5	52.7	57.4
Intrastate permits[b]	32.4	40.6	52.5
Interstate permits[c]	30.5	38.8	51.4
Emission taxes[d]	29.8	38.8	49.5
	Investment in clean-coal plants combined (GW)		
Performance standard[a]	9.3	18.6	34.9
Intrastate permits[b]	9.6	18.6	27.6
Interstate permits[c]	9.2	19.4	27.5
Emission taxes[d]	10.1	19.3	27.7

[a] Emissions from every plant required to be less than or equal to 2.0 lbs/mmBtu, equivalent to a national 10 mt/year emissions reduction target.

[b] 10 mt annual emissions reductions specified at the state level through issue of permits.

[c] 10 mt annual emissions reductions specified at the regional level through issue of permits.

[d] Taxes set so that in the expected future, annual emissions are reduced by 10 mt per year, by the year 2000.

The bottom half of table 6-1 indicates that, unlike FGD retrofits, investments in new clean-coal generating capacity are not greatly affected by the environmental policy design except in the upper tails of the distributions, where the rigid standard again requires more investment. This finding echoes our results in chapter 5, where we found in the sensitivity analysis that the dominant influences on new investment were related to demand (see table 5-9). This conclusion is sustained by formal sensitivity analysis of our model results for different environmental policy designs, using the partial rank correlation methods described in earlier chapters. For brevity, we do not report the full results here. We also find that for each policy design the most significant influence on electricity cost is demand growth (see also chapter 5, and table 5-11).

Where the policy design does seem to make a difference in the results is in the importance of IGCC cost as an ultimate influence on electricity costs. This factor appears to be more important under a permit trading approach—where there is greater latitude for compliance—than under a command-and-control approach. This result illustrates how the flexibility accorded utilities under incentive-based environmental policies expands the dimensions of their decision planning problems.

7

DIRECTIONS FOR FURTHER RESEARCH

In this study we have described a methodology for discerning economic cost-minimizing technology choices for electricity generation, and the influences most strongly affecting these choices, under different assumptions about costs, demand, and environment regulation. This analytical structure can shed light on cost-minimizing technology choices over time for several geographical regions, while including a variety of different assumptions about interregional trade. Perhaps the most important and distinctive feature of the methodology is its ability to provide a fairly rich description of the pattern of how technology choices may vary with changes in assumptions about future economic and regulatory influences. The flexibility of the model allows us to reflect diverse points of view about controversial issues in an effort to seek general conclusions.

As pointed out often in the course of the study, the results of our modeling analysis—like any other such exercise—are affected by limitations in both the model structure and input assumptions. We believe there is enough of interest in the results to warrant further investigation of the issues addressed in the study. An agenda for extending our research is sketched below. Some extensions, for example those involving different input assumptions, would be routine. Others involving certain changes in the model structure would entail considerably more analytical effort.

Further experiments with different input assumptions. We have noted earlier that the findings of our analysis appear to be significantly shaped by our assumptions (drawn partly from EPRI studies) concerning the incremental costs and performance of alternative technologies. The relative economics of different advanced technologies remain controversial. This issue could be addressed in part by running the model with the same slate of technologies we have considered here, but with additional cost and performance assumptions. A wider range of fuel-price escalation rates could also be included in this extension, as well as a more detailed analysis of the impacts of different load-management programs. In general, the analysis could be strengthened by introducing more correlations among input parameters in the sensitivity analysis (for example, associating higher gas prices with faster growth in electricity demand), rather than treating the parameters as statistically independent in the sampling.

Changes in the menu of technologies assumed to be available for new capacity investment. Changes in the technology menu to incorporate dispatchable cogeneration technologies, utility investment in conservation programs, and endogenously determined plant-life-extension investments would also be desirable. However, the first two of these extensions would entail considerable research to develop the necessary assumptions about input parameters, while the third would require significant alterations of the existing model or development of a new one. Changes in the technology menu could also reflect the choice of a longer time horizon than 2010, in which additional technologies could be viewed as commercially feasible, or the choice of possible investment in a new generation of nuclear plants.

Expansion of the geographical scope of the model. The analysis in this study addresses only one nine-state area. To provide a more solid basis for policy recommendations, the analysis must be national in scope and must be based upon the most up-to-date, accurate assumptions about demand, existing plants, and so forth. Broadening the application of the model to a national scope requires further investigation of interregional trading patterns. The most in-depth specification of inputs would require access to proprietary data.

Inclusion of other environmental constraints on power production, notably limits on nitrogen oxides (NO_x) and carbon dioxide (CO_2), and further analysis of alternative environmental policies. A fuller treatment of issues related to acid rain precursors requires the inclusion of constraints on NO_x as well as SO_2 emissions. In addition, hazardous waste disposal should be factored into the analysis. Addressing ''greenhouse'' issues would require a different set of assumptions about technology performance and an expansion of the technology menu for both electricity generation and pollution control. As we found in chapter 6, a method of environmental control can have effects beyond those that result from the stringency of the total emission standard. Further analysis of environmental means and ends would help to clarify the impacts of environmental costing proposals that require utilities to include externalities in their planning decisions.

Extension of the analytical framework to address transmission issues. In modeling interstate power trade in this study, we have assumed either that no new power exchanges are feasible or that any power exchange will be undertaken whose gross trading value exceeds the long-run marginal cost of transmission. Clearly other intermediate possibilities could be considered, including different types of physical limits on transmission capacity and different specifications for the sharing of cost savings. A more ambitious undertaking still would entail extending the structure of the model to include transmission investment along with generation investment as a model output.

All of these extensions would deepen our understanding of technology choices that minimize the economic cost of electricity supply, taking into account market conditions and environmental constraints. Of course, as we noted at the outset of the book, such knowledge by no means ensures that economic cost-minimizing choices will be undertaken; nor does it necessarily follow that these cost-minimizing choices should be undertaken. In this study we have abstracted from the complex structure and consequences of economic regulation in the electricity industry. A better understanding of barriers to economic efficiency, and of means by which incentives for efficient performance can be strengthened, obviously would be a vital complement to the subjects we have addressed.

REFERENCES

Baughman, M. L., P. L. Joskow, and D. P. Kamat. 1979. *Electric Power in the United States: Models and Policy Analysis* (Cambridge, Mass., MIT Press).

Bohi, D. R., D. Burtraw, A. J. Krupnick, and C. G. Stalon. 1990. "Emissions Trading in the Electric Utility Industry," Discussion Paper QE90-15, Quality of the Environment Division, Resources for the Future (Washington, D.C.).

Bombaugh, K. J., and W. J. Rhodes. 1988. "Discharges from Coal Gasification Plants," *Environmental Science and Technology* vol. 22, no. 12 (December), pp. 1389–1396.

Cicchetti, C. J., and W. W. Hogan. 1988. "Including Unbundled Demand Side Options in Electric Utility Bidding Programs," Energy and Environmental Policy Center Discussion Paper E-88-07, Harvard University (August).

Cushey, M. A., and E. S. Rubin. 1988. "Simplified Models of U.S. Acid Rain Control Costs," *Journal of the Air Pollution Control Association* vol. 38, no. 12 (December), pp. 1523–1527.

Dantzig, G. B., and P. Wolfe. 1960. "Decomposition Principle for Linear Programs," *Operations Research* vol. 8, no. 1 (January-February), pp. 101–111.

Data Resources Incorporated (DRI). 1987. *Energy Review* (Winter).

Department of Energy (DOE). 1987a. *America's Clean Coal Commitment.* Document DOE/FE-0083, Office of Fossil Energy (Washington, D.C., February).

———. 1987b. *Second Report to Congress on Emerging Clean Coal Technologies Capable of Retrofitting, Repowering, Modernizing Existing Facilities.* Document DOE/FE-0086 (Washington, D.C., May).

———. 1988. *Electric Power Supply and Demand for the Contiguous United States, 1987–1996.* Document DOE/IE-008 (Washington, D.C., February).

Dowlatabadi, H. 1984. "Electricity Interchange Between Integrated Grid Systems: Methods and Case Studies" (Ph.D. dissertation, Energy Research Group, Cambridge University).

Dowlatabadi, H., and N. Evans. 1986. "Electricity Trade in the U.K.: Economic Prospects and Future Uncertainty," *Energy Policy* vol. 14, no. 1 (February), pp. 35–47.

Dowlatabadi, H., and W. Harrington. 1989. "The Effects of Uncertainty on Policy Instruments: The Case of Electricity Supply and Environmental Regulation," Discussion Paper QE89-20, Quality of the Environment Division, Resources for the Future (Washington, D.C.).

————, and ————. 1990. "Uncertainty and the Cost of Acid Rain Control," *Contemporary Policy Issues* vol. 8, no. 3 (July), pp. 43–58.

Dowlatabadi, H., R. Edahl, N. Tyle, S. N. Talukdare, and N. L. Pachavis. 1986a. *Padre Analytical Documentation* (Pittsburgh, Pa., Center for Energy and Environmental Studies, Carnegie Mellon University).

————. 1986b. *Padre Design Documentation* (Pittsburgh, Pa., Center for Energy and Environmental Studies, Carnegie Mellon University).

————. 1986c. *Padre Programming Documentation* (Pittsburgh, Pa., Center for Energy and Environmental Studies, Carnegie Mellon University).

Downie, N. M., and R. W. Heath. 1965. *Basic Statistical Methods* (2d ed., New York, Harper and Row).

Edison Electric Institute (EEI). 1987. *1985 Capacity and Generation of Non-Utility Sources of Energy* (Washington, D.C., April).

Electric Power Research Institute (EPRI). 1979. *Uncertainty Methods in Comparing Power Plants.* Report FFAS-1048 (Palo Alto, Calif., April).

————. 1984. *Effects of Resource Depletion on Future Coal Prices.* Report EA-3733 (Palo Alto, Calif., October).

————. 1986a. *Impact of Demand Side Management on Future Customer Electricity Demand.* Report EM-4815-SR (Palo Alto, Calif., October).

————. 1986b. *Technical Assessment Guide,* vol. 1: *Electricity Supply 1986.* Report P-4463-SR (Palo Alto, Calif., December).

Energy Information Administration (EIA). 1986. *Historical Plant Cost and Annual Production Expenses for Selected Power Plants, 1984.* Document DOE/EIA-0455 (Washington, D.C., Government Printing Office).

————. 1988a. *Annual Energy Outlook 1987.* Document DOE/EIA-0383 (87) (Washington, D.C., Government Printing Office, March).

————. 1988b. *Historical Plant Cost and Annual Production Expenses for Selected Power Plants, 1986.* (Washington, D.C., Government Printing Office).

Energy Ventures Analysis, Inc. 1985. "Evaluations of SO_2 Emissions and the FGD Retrofit Feasibility at the 200 Top Emitting Generation Stations." Report prepared for the U.S. Environmental Protection Agency (Washington, D.C., April).

Federal Energy Regulatory Commission (FERC). 1987. "Generic Determination of Rate of Return on Common Equity for Utilities," Docket no. RM87-35-000.

Hogan, W. W. 1988. "Oil Demand and OPEC's Recovery," Energy and Environmental Policy Center Discussion Paper E-88-02, Harvard University (June).

Iman, R. L., and J. C. Helton. 1985. *A Comparison of Uncertainty and Sensitivity Analysis Techniques for Computer Models.* Report NOREG/CR-3904 SAND84-1461 (Albuquerque, N.M., Sandia National Laboratories, March).

Joskow, P. L. 1987. "Productivity Growth and Technical Change in the Generation of Electricity," *Energy Journal* vol. 8, no. 1 (January), pp. 17–38.

———— 1989. "Regulatory Failure, Regulatory Reform, and Structural Change in the Electrical Power Industry," in M. N. Baily and C. Winston, eds., *Brookings Papers on Economic Activity: Microeconomics 1989* (Washington, D.C., Brookings Institution).

Joskow, P. L., and N. L. Rose. 1985. "The Effects of Technical Change, Experience, and Environmental Regulation on the Construction Cost of Coal-Burning Generating Units," *Rand Journal of Economics* vol. 16, no. 1 (Spring), pp. 1–27.

Joskow, P. L., and R. Schmalensee. 1983. *Markets for Power* (Cambridge, Mass., MIT Press).

————, and ————. 1985. "The Performance of Coal-Burning Electric Generation Units in the United States: 1960–1980," Department of Economics Working Paper no. 377, Massachusetts Institute of Technology (July).

Lasdon, L. S. 1966. "Duality and Decomposition in Mathematical Programming," *IEEE Transactions on System Science and Cybernetics* vol. SSC-4, pp. 36–40.

————. 1970. *Optimization Theory for Large Systems* (New York, Macmillan).

Lehman, E. L. 1975. *Nonparametrics: Statistical Methods Based on Ranks* (Oakland, Calif., Holden-Day).

Lind, R. C., ed. 1982. *Discounting for Time and Risk in Energy Policy* (Washington, D.C., Resources for the Future).

Massé, P., and R. Gilbert. 1964. "Application of Linear Programming to Investments in the Electric Power Industry," in R. J. Nelson, ed., *Marginal Cost Pricing in Practice* (Englewood Cliffs, N.J., Prentice-Hall).

Merrow, E., K. Phillips, and C. Myers. 1981. *Understanding Cost Growth and Performance Shortfalls in Pioneer Process Plants* (Santa Monica, Calif., Rand Corporation).

North American Electricity Reliability Council (NERC). 1987. *1987 Electricity Supply and Demand: for 1987–1996* (Princeton, N.J., November).

Rose, N. L., and P. L. Joskow. 1988. "The Diffusion of New Technologies: Evidence from the Electric Utility Industry," Working Paper no. 2676, National Bureau of Economic Research (August).

Rothwell, G. S. 1986. "Steam-Electric Scale Economies and Construction Lead Times," Social Science Working Paper 627, Division of the Humanities and Social Sciences, California Institute of Technology (December).

Rubin, E. S. 1983. "International Pollution Control Costs of Coal-Fired Power Plants," *Environmental Science and Technology* vol. 17, no. 8 (August), pp. 366A–377A.

Tietenberg, T. H. 1985. *Emissions Trading: An Exercise in Reforming Pollution Policy* (Washington, D.C., Resources for the Future).

Toman, M. A., and J. Darmstadter. 1988. "Improving Performance of Wholesale Electric Generation Markets," Discussion Paper ENR88-03, Energy and Natural Resources Division, Resources for the Future (Washington, D.C.).

Turvey, R. 1968. *Optimal Pricing and Investment in Electricity Supply* (Cambridge, Mass., MIT Press).

Turvey, R., and D. Anderson. 1977. *Electricity Economics* (Baltimore, Md., The Johns Hopkins University Press for the World Bank).

Wismer, D. A., ed. 1971. *Optimization Methods for Large-Scale Systems* (New York, McGraw-Hill).

Wolk, R., and N. Hold. 1988. "The Environmental Performance of Integrated Gasification Combined-Cycle (IGCC) Systems," paper presented at the Fourth Symposium on Integrated Environmental Control, Washington, D.C. (Palo Alto, Calif., Electric Power Research Institute, March).

ABOUT THE AUTHORS

Hadi Dowlatabadi is an associate professor in the Department of Engineering and Public Policy, Carnegie Mellon University, and a former fellow in the Energy and Natural Resources and the Quality of the Environment divisions at Resources for the Future. Since receiving his Ph.D. in physics from Cambridge University in 1984, his research has focused on technological and environmental issues confronting the electricity industry in the United States and the United Kingdom. He has coauthored several scholarly papers on these issues, and has served on advisory panels to the U.S. Environmental Protection Agency and other organizations.

Michael A. Toman is a senior fellow in the Energy and Natural Resources Division at Resources for the Future and a professorial lecturer at the Johns Hopkins School of Advanced International Studies. He received his Ph.D. in economics from the University of Rochester in 1983. He is coauthor of the RFF book *Analyzing Nonrenewable Resource Supply,* and has written numerous papers on a wide range of topics in resource and regulatory economics.